Epoxy-Based Composites

Edited by Samson Jerold Samuel Chelladurai, Ramesh Arthanari and M.R.Meera

Published in London, United Kingdom

Epoxy-Based Composites
http://dx.doi.org/10.5772/intechopen.97898
Edited by Samson Jerold Samuel Chelladurai, Ramesh Arthanari and M.R.Meera

Contributors
Jun Zhang, Hai Luo, Xiaojian Zhou, Bowen Liu, Zade S Kiran, Syam Kumar Chokka, V. Suresh Babu, K.V. Sai Srinadh, Xianhui Zhao, Shun Lu, Wei Li, Shuyang Zhang, Kai Li, Kashif Nawaz, Peter Wang, Guang Yang, Arthur Ragauskas, Soydan Ozcan, Erin Webb, Hayriye Hale Aygün, Nassima Radouane, Abdelkrim Maaroufi, Srikanta Moharana, Bibhuti Bhusan Sahu

First published in London, United Kingdom, 2022 by IntechOpen
IntechOpen is the global imprint of INTECHOPEN LIMITED, registered in England and Wales, registration number: 11086078, 5 Princes Gate Court, London, SW7 2QJ, United Kingdom

British Library Cataloguing-in-Publication Data
A catalogue record for this book is available from the British Library

Additional hard and PDF copies can be obtained from orders@intechopen.com

Epoxy-Based Composites
Edited by Samson Jerold Samuel Chelladurai, Ramesh Arthanari and M.R.Meera
p. cm.
Print ISBN 978-1-80355-159-3
Online ISBN 978-1-80355-160-9
eBook (PDF) ISBN 978-1-80355-161-6

For EU product safety concerns:
IN TECH d.o.o., Prolaz Marije Krucifikse Kozulić 3, 51000 Rijeka, Croatia,
info@intechopen.com or visit our website at intechopen.com.

Meet the editors

Dr. Samson Jerold Samuel Chelladurai obtained his BE in Mechanical Engineering, ME in Manufacturing Engineering, and Ph.D. in Composite Materials from Anna University, India. He has published many research articles in SCI and Scopus-indexed journals. He has also published five books and three patents and received grants worth $89,000 USD. He has received several awards including grants, Best Research Paper, Best Faculty, Star Performer, and Shri. P. K. Das Memorial Best Faculty Award in Mechanical Engineering, Best Academic Non-Circuit Faculty, and Young Researcher Award in Mechanical Engineering from Sri Krishna College of Engineering & Technology (SKCET) and other organizations.

Dr. Ramesh Arthanari is working as a Principal at the Chennai Institute of Technology, India. He obtained a BE in Mechanical Engineering from Government Engineering College, Salem, India, and an ME from Government College of Technology, Coimbatore, India. He completed a Ph.D. in Metal Casting in 2007. He has published many technical papers in national and international journals. He has received around $50,000 USD for establishing a research center for composites. He has published three patents and has received several awards from reputed professional organizations such as the Society of Automotive Engineers (SAE) and other institutions. His areas of interest include composite materials and tribology.

Dr. M.R. Meera is an assistant professor in the Department of Physics, Sree Ayyappa College for Women, India. She has a BSc, MSc, and MPhil in Physics. She also has a BEd, MEd, and MPhil in Education. She secured the university's first rank with a Gold Medal for outstanding academic performance in her MPhil and fifth rank in her MEd She obtained a Ph.D. in Material Science from Manonmanium Sundaranar University, India, in 2018. She completed a UGC minor research project in 2011. She is currently guiding four Ph.D., research scholars. She has also published several papers in reputed international journals.

Contents

Preface

Epoxy-based composites are widely used in aerospace, automotive, marine, and defense applications because of their excellent mechanical properties and good resistance to wear and corrosion. Polymer matrix composites can be prepared using the hand lay-up technique, compression molding, and spray-up and pultrusion processes.

This book presents the use of epoxy as filler for polymer matrix composites, synthesis and properties of epoxy-based composites, fiber inclusion of epoxy-based composites, research work on epoxy-glass composites, modifications of epoxy resin adhesives, and application of epoxy-based composites for radiation shielding.

This book is useful for industrialists, undergraduate and postgraduate students, research scholars, and scientists to understand the use of epoxy-based composites, manufacturing methods, mechanical testing, and applications of epoxy-based composites in various sectors.

Dr. Samson Jerold Samuel Chelladurai
Associate Professor,
Department of Mechanical Engineering,
Sri Krishna College of Engineering and Technology,
Coimbatore, India

Dr. Ramesh Arthanari
Principal and Professor,
Chennai Institute of Technology,
Chennai, India

Dr. M.R. Meera
Assistant Professor,
Department of Physics,
Sree Ayyappa College for Women,
Kanyakumari, India

Section 1

Resins

Chapter 1

Epoxy as Filler or Matrix for Polymer Composites

Xianhui Zhao, Shun Lu, Wei Li, Shuyang Zhang, Kai Li, Kashif Nawaz, Peter Wang, Guang Yang, Arthur Ragauskas, Soydan Ozcan and Erin Webb

Abstract

Epoxy is a widely used polymer because of its ease of processing, high adhesiveness, and high chemical resistance. Epoxy-based composites are commonly used in aerospace, automotive, and marine applications. The epoxy type, function, curing agent, and curing process are discussed in this chapter. Epoxy is used as either a filler or polymer matrix in composite applications. As a filler, the epoxy modification on the fiber is discussed. As a polymer matrix, the epoxy is reinforced by natural and synthetic fibers. The manufacturing process and the fabricated epoxy-based composites' performance (e.g., mechanical and thermal properties) are investigated. The advantages and disadvantages of epoxy's function are discussed and summarized. Epoxy modification is an effective approach to improve the composites' performance.

Keywords: epoxy, filler, fiber, matrix, polymer, composite

1. Introduction

Since the discovery of epoxy in 1909 by Prileschajew [1], epoxy has become an important part of our daily life and integral to numerous industrial sectors. Epoxy is a plastic that has an epoxy ring, which consists of two carbon atoms that are bonded to a common oxygen atom. Epoxy resins are a class of reactive prepolymers and polymers that contain epoxide groups, which are cured using a wide variety of curing agents (i.e., hardener). Epoxy is one of the most important thermosetting polymers and has wide applications in different areas. The products' properties largely depend on the structure of the epoxy resins and curing agents used. The excellent mechanical properties, high adhesiveness to many substrates, low shrinkage after curing, and good heat and chemical resistances make epoxy resins remarkably useful across a wide range of fields, where they act as reinforced materials, adhesives, coatings, etc. [2–5].

Epoxy has been widely used as polymer matrix for composite applications. The properties of composites can be customized by adding functional fibers to endow the functionalities on the composites. For example, the fiber-reinforced epoxy composites have significantly improved mechanical properties. Song et al. [6] developed carbon fiber/epoxy composite laminates with carbon nitride, and their tensile strength and Young's modulus reached 67 MPa and 58 GPa, respectively. Huang et al. [7] developed continuous bamboo fiber reinforced epoxy composites (i.e., long bamboo fiber as reinforcement rather than chopped fibers through a resin transfer molding process and achieved improved mechanical strain. Epoxy based composites not only have enhanced mechanical properties, they can also feature unique properties, such as electromagnetic interference shielding properties [8], self-healing [9], heat resistance [10], and others [3, 4].

Despite serving as a polymer matrix, epoxy, as a reactive molecule, can be used as fillers (or modifiers) to facilitate reinforce polymer composites. Fillers are typically added to polymers to modify their characteristics and/or reduce their price [11–13]. For instance, Zhao et al. applied epoxy to functionalize pine fibers that were used to reinforce the polylactic acid (PLA) [11]. Their results suggest that the epoxy modification is beneficial to improve the tensile strengths and Young's moduli of the composites compared to that of neat PLA owing to the improved fiber/matrix interfacial adhesion. One reason is that the epoxy exhibits high adhesiveness and compatibilization effect. In another study conducted by Sujaritjun et al. [12], epoxidized polybutadiene treated bamboo fibers were used to reinforce PLA and achieved a 10% increase in the tensile strength. Similarly, Kyutoku et al. [13] utilized cellulose fibers coated with epoxy-based agents to reinforce PLA and found that the epoxy treatment improved the interfacial adhesion between cellulose fibers and PLA. Compared with other non-reactive surface functionalization of the nanofiber [14], the high reactivity of the epoxy group in the epoxy modified fibers is beneficial for improving the interfacial interactions between fibers and matrix, and thus can effectively reinforce polymers.

The cured epoxy resins are brittle due to the high degree of cross-linking, which weakens epoxy's impact strength and other relevant properties [3, 15]. Therefore, chemical modification of epoxy monomers is necessary to improve their flexibility and toughness as well as other properties [16]. Many efforts have been devoted to addressing these challenges through introducing flexible polymers, inorganic solid particles, and elastomers [17–19]. Despite the brittleness, epoxies also have poor thermal and electrical conductivity [20]. To solve these challenges and improve the properties of the epoxy-based materials, different thermal conductive and electrical conductive materials (fibers), such as carbon nanotubes, graphene, carbon fiber, and others, were added [21]. However, the poor interfacial interaction between the fiber and epoxy led to low interfacial strength and impact toughness. Considerable efforts have been made to solve this issue [6, 7]. Additionally, bio-based epoxy [22] and epoxy-based vitrimer [23] also become new hot topics in this field driven by the desire for sustainability and recyclability [24] of composite materials.

Given the importance of the epoxy-based composites, in this chapter, the recent development of epoxies in polymer composites are summarized based on their roles. The basics of epoxy, such as structure, curing agents, and curing process, are discussed. The role of the epoxy compounds in the composites is discussed from the filler and polymer matrix aspects. Finally, the challenges and future opportunities in the epoxy-based composites are summarized.

2. Epoxy

2.1 Epoxy type

Epoxy resin is often regarded as the most valuable polymer. It is widely utilized in numerous modern industries including aerospace [25], electrical encapsulation [26], chemical engineering [27], civil engineering [28], food industry [29], etc. Epoxy resins are part of the family of monomeric or oligomeric materials with excellent electrical insulation properties and extraordinary resistance to corrosion [30]. Epoxy resin has a group of reactive polymers, which contain epoxy or oxirane groups, as presented in **Figure 1**. According to their structure, characteristic, and functionality, there are three major types of epoxy resins: (i) cycloaliphatic, (ii) epoxidized, and (iii) glycated epoxy resins [31].

The first two types of resins (cycloaliphatic and epoxidized) are synthesized directly via the corresponding olefin oxidation. Glycidated resins are prepared by glycidation of bisphenol A with epichlorhydrin. Generally, cycloaliphatic epoxy resins provide excellent ultraviolet (UV) stability and electrical properties, as well as good thermal stability due to their fully saturated structure. This feature makes them suitable for structural components under high-temperature situations. Furthermore, these resins are easy to obtain owing to their low viscosities and can be cured at low temperatures. This result is likely because these resins have simple structures, and the low viscosity leads to easy curing. However, they have a short pot life (useable life: the length of time that a system retains a viscosity low enough to be applied), as moisture and other factors easily affect the uncured resins. Bisphenol epoxy resin is a type of commercial resin, commonly composed of a mixture of oligomers. It has a wide variety of viscosities and low molecular weights, resulting in high-performance phenoxy lacquer resins [31].

2.2 Epoxy functions and advantages

Epoxy resins are utilized in various applications due to their outstanding properties and versatile nature [25, 29, 31]. Epoxy resins can form into thermoset or photosensitive resins, which are determined by reaction conditions and curing agents [32, 33]. Epoxy-based parts are usually manufactured in the desired shape by typical technologies including but not limited to resin infusion [34], injection molding [35], resin transfer molding [36], and autoclave [37]. Nevertheless, the above processed technologies cause several bottlenecks in terms of slow production, difficulties in producing complex geometries, labor-intensiveness, and high cost of the process, particularly considering thermal energy [31].

One significant characteristic of epoxy resins is that the curing process proceeds through a step-growth reaction without the generation of volatile byproducts. Subsequently, the process produces structures free from voids and bubbles. Therefore, the tailoring of cross-linkers and modifiers are critical in the steady growth rate of epoxy resins considering their further functionality. Because of the poor toughness,

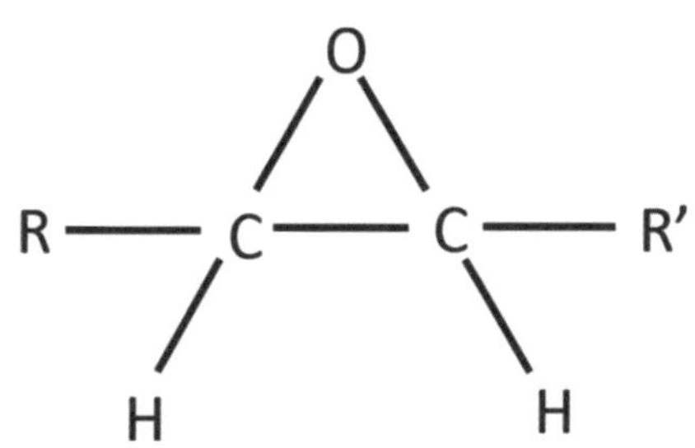

Figure 1.
Epoxy or oxirane group.

researchers explored and developed several strategies to improve the toughness of epoxy resins: (i) change in the chemical structure to make them more flexible [38], (ii) increase in the molecular weight [39], (iii) decrease in the cross-linking density [40], (iv) addition of a tougher phase, and (v) addition of fibers [29].

2.3 Epoxy curing agent

The desired properties of epoxy resins are generally determined by combining an epoxy resin and a curing agent. Curing agents can be catalysts usually drawn from amines, Lewis acids, or hardeners. Hardeners, identified as curing or cross-linking agents, can break the C-O-C ring at the ends of an epoxy molecule and attach themselves to the molecules to convert the resin into a thermoset network structure. Different curing agents can create composites with different properties because the curing agents have different functional sites that lead to different chemical reactions. A specific hardener should be selected based on processing conditions (viscosity, pot life, mixing ratio, and temperature) and the desired properties of the product (strength, chemical and thermal resistance, toughness, and flexibility). The most common hardeners for epoxy resins are amine types: (i) aliphatic, (ii) aromatic, and (iii) cycloaliphatic. The type, amount, functionality of the hardener, as well as the curing conditions (temperature and curing time), affect the properties of hardened epoxy resins. For example, difunctional epoxy resins are prepolymers with an epoxy group at each end of a molecule. The diglycidyl ether of bisphenol F (DGEBF) has similar molecular structures: the two methyl groups attached to the carbon between the benzene groups in the former are replaced with H atoms in the latter. F-type epoxies have lower viscosity than A-type epoxies. For this reason, A and F types are often mixed to lower the viscosity of a mixture, thus preventing the crystallization of the polymers [41].

Ozkul et al. [41] employed three different epoxy resins, diglycidyl ether of bisphenol A, diglycidyl ether of bisphenol F (DGEBF), and a mixture of the two; six amine type hardeners, four of which are aliphatic and the others are cycloaliphatic; and three glycidyl-ether based reactive diluents to study their effects towards the hardened state properties of epoxy mortars. Three epoxy resins and three types of glycidyl ether-based reactive diluents were used together in epoxy mortars. Six different amine-based hardeners, four aliphatic and two cycloaliphatic, were used in the mixtures. A hardener with a higher functionality (reactive point) led to mortars with higher strength.

2.4 Epoxy curing process

Epoxy resins are in liquid form and the curing agent solidifies during the fabrication. During curing, the chemical reaction of the epoxide groups in the epoxy resin is initiated by a curing agent through a step-growth reaction, then form a 3D highly cross-linked network. Curing process conditions impact the properties of the cured resins. For example, the amount of curing agents influences the strength of the cured resin film and the formation time. There are three main methods of curing: (i) room-temperature curing, (ii) thermal curing, and (iii) photocuring [31].

Room temperature curing agents are used to cure epoxy resins at room temperature. These include aliphatic polyamines, low-molecular-weight polyamides, alicyclic polyamines, and modified aromatic amines. This curing system is preferred for higher flexibility, impact resistance, and electrical and thermal shock resistance [31]. The final structure of the cured resin highly depends on the employed temperature.

High temperature curing agents, for example, acid anhydrides, resol resins, aromatic polyamines, dicyandiamides, and hydrazides, are used to cure epoxy resins at elevated temperatures. Generally, high temperature curing is carried out in two stages: (i) precuring stage (gel-status) at a low temperature and (ii) postcuring stage (solid-status) at a high temperature. Epoxy components fabricated at higher temperatures exhibit higher tensile strength, heat resistance, and chemical resistance [31].

Photocuring systems significantly decrease the epoxy curing process time from hours to minutes and save the cost of energy compared to autoclave thermal curing. A wide range of radiation curing wavelengths, such as infrared light, UV light, X-rays, and electron beam irradiations, can be used to cure photosensitive epoxy resins employing photoinitiators. Photocuring is a relatively controllable and consistent process as compared to other curing processes [31].

Two-stage curing systems that combine photocuring and thermal curing enable the use of epoxy resins in additive manufacturing. Traditionally, epoxy resin and its composites are manufactured by curing at high temperatures for several hours. This condition makes epoxy resin impossible to use in additive manufacturing because, curing layer by layer is essential during fabrication. To solve this issue in additive manufacturing the common practice is to use photosensitive epoxy resin. However, this cured resin has poor mechanical properties, low curing rate, and high cost due to photoinitiators. Moreover, it does not meet the need for mechanical parts required in some specified industries. Direct-ink writing is a possible approach to use epoxy or other thermoset resins where the ink can be designed to be viscoelastic with the incorporation of a fiber [31].

Kuang et al. [42] demonstrates additive manufacturing of an epoxy resin using digital light processing. The hybrid ink is prepared by blending a thermally curable resin and a photocurable resin and cures in two stages. During manufacturing, the epoxy resin is exposed to UV-LED light with a wavelength of 385 nm (first-stage curing) to create the designed shape and then cured at 100°C for 2 h and another 2 h at 160°C (second-stage curing). The hybrid ink presents low volume shrinkage, desired mechanical properties and high resolution, enabling use in engineering applications. Griffini et al. [43] customizes a hybrid ink of thermal and photocurable epoxy resins to investigate the possibility of the fabrication of carbon fiber-reinforced composite structures. Carbon fibers with a diameter of 7.2 μm and length of 100–150 μm are added into the ink to develop an increased weight concentration. When the hybrid ink is extruded from the syringe, light from two UV-A torches with a wavelength of 405 nm irradiates the hybrid ink to solidify its shape in the first-stage curing. Then, the printed sample is cured at 220°C for 20 min in a ventilated oven for the second-stage curing. The weight ratio between thermal and photocurable components strongly influenced the printing characteristics. Carbon fibers also affected the efficiency of UV-curing.

3. Epoxy-based composites

3.1 Epoxy as filler

Epoxy can be used as a non-matrix component (i.e., filler or minor component) in composites and is usually applied at the fiber and polymer matrix interface. Epoxy can be used to functionalize the surface of fibers with epoxy groups. Thus, the surface of the fibers can be activated to react with other components in the composites.

Cheng et al. [44] uses a two-step method to epoxidize the surface of poly-p-phenylene-benzimidazole-terephthalamide (PBIA) fibers with epoxy groups. The PBIA fibers were first fluorinated and grafted with Si-OH groups. The Si-OH groups were then added with 3-glycidoxypropylthrimethoxysilane (GPTMS) for epoxidation. The modified fibers were then incorporated in an epoxy matrix with respect to surface energy and pull-out strength, where both the two properties were improved. The surface energy increased from 13 to 17 mN/m, while the pull-out strength improved from 0.45 to 0.63 N/(tex·mm). Fei et al. [45] used the epoxy group as an activate site to graft polymer chains for functional composites. The bacterial cellulose was first epoxidized with epoxy chloropropane (ECP) at the hydroxyl groups on cellulose. Aniline was then added and grew from the epoxy group to polyaniline (PANI), providing electrical conductivity in the final cellulose-based flexible paper. The electrical conductivity can reach 1.1 S/cm at proper ECP concentrations and can be further improved up to 1.4 S/cm by introducing polyacrylamide via enhancing the interaction between PANI and the fibers.

Another epoxidation was done by Cao et al. [46] on tunicate cellulose nanocrystals (T-CNC), which exhibit a higher aspect ratio than cotton CNC. GPTMS was used to introduce epoxy groups on the surface of T-CNC first. The modified fibers were then blended with carboxylate styrene–butadiene rubber. The covalent bonding between epoxy and carboxyl groups was improved. The tensile strength, modulus, and toughness were improved with more modified T-CNC added. In addition, the resultant rubber could be reprocessed, attributed to the transesterification of the ester bonding formed by epoxy and carboxyl groups. This modification enabled the composites to maintain their original tensile strength even after reprocessing three times. In these methods, epoxy significantly altered the surface chemistry and the interfacial adhesion. However, one issue was that the epoxidation process usually involved highly activate reagents, like ECP, which may bring an uncontrollable by-reaction that will affect the properties of the final product [45].

Epoxy are utilized to adjust the properties of composites. A single domain or phase with an epoxy group is introduced into composites. The properties of the composites change depending on the inherent properties of the epoxy group. Epoxy can also react with other components to bring new bonding. For example, Immonen et al. [47] used epoxidized linseed oil (ELO) as an epoxy phase in a PLA-based composites with bleached softwood kraft pulp (BSKP) as the fiber. At certain concentrations of ELO, the trinary composites exhibited improved tensile strength, modulus, elongation at break, and impact strength compared with neat PLA and BSKP/PLA composites. This result inferred that ELO could react with both PLA and BKSP, thus enhancing the interfacial adhesion between PLA and BSKP. A similar idea was applied by using tannic acid-crosslinked epoxidized soybean oil (TA-ESO) in bamboo fiber (BF)/PLA composites as reported in Liu et al.'s work [48]. In their composite system, BF was the reinforcing fiber that could improve tensile strength and modulus as more incorporated, while decreasing the elongation at break. TA-ESO was introduced by solution-spray onto the BFs and then dried to modify the surface of BF for better interaction with the PLA matrix. TA-ESO was expected to act as a bridge between the two phases to enhance the interfacial adhesion. At a low concentration (0.5 wt%) of TA-ESO, the epoxy phase mainly acted as a compatibilizer between PLA and BF since hydroxyl groups on both PLA and BF could react with epoxy group on EOS, contributing to the force transfer from PLA to BF to improve the tensile strength. As TA-EOS content increased, excessive TA-EOS could be presented as a single phase, which acted as a toughing domain in PLA composites and a compatibilization effect on the interface.

However, TA-EOS at higher concentration resulted in poor tensile performance and impact strength due to the large domain area. Nevertheless, TA-EOS could be used to balance the tensile strength and toughness by varying the feeding ratio compared to BF and PLA matrix.

Similarly, Zhao et al. [11] impregnated an epoxy solution to pine fibers to fill and penetrate the voids and hollow channels in the pine fibers. By mixing curing agent (Dicyandiamide) and epoxy resin (Poly(bisphenol A-co-epichlorohydrin), glycidyl end-capped), part of the epoxy cured during the drying process, modifying the surface chemistry. This method benefited the tensile strength, where the tensile strength and Young's modulus improved by 20% and 82%, respectively, at 1 wt% loading of the epoxy system. The better interfacial adhesion between PLA and pine fibers came from the compatibilization effect of the epoxy. Epoxy curing during the hot pressing for final composites may bridge PLA and pine fibers by a covalent bond. The epoxy enhanced the interaction between polymer matrix and pine fibers increasing the tensile performance. Through this approach, the preparation of the composites was conducted without complex chemical reactions. However, the property of the final products depended on the inherent property and concentration of the epoxy-included phase which should be carefully selected [48].

3.2 Epoxy as polymer matrix

Epoxies are extensively used as the polymer matrix of composites for their salient features of high specific strength, good dimensional stability, and excellent dielectric properties [1, 30, 49]. Due to the high cross-link density formed during curing, neat epoxy resins are intrinsically brittle and have low fracture toughness, which can limit their application in some specified fields. Moreover, epoxy resins are relatively expensive in comparison with other commonly used thermosetting resins, such as unsaturated polyester and vinyl ester resins. Adding fibers into epoxies could enhance the thermal and durability properties [50]. However, the addition of fibers can impart adverse influence on composites' physical and mechanical properties alongside the reinforcement on other properties. Therefore, attention needs to be paid to optimizing and balancing different properties and costs when designing composites for specific applications. **Table 1** summarizes the general types and fabrication techniques of epoxy-based composites. Some applications and recent developments of high-performance and cost-effective epoxy composites are briefly reviewed herein.

Fiber reinforcement is a common strategy for preparing epoxy composites and is widely adopted in the industry [50, 54]. Synthetic fibers, such as glass, carbon, and aramid, offer load bearing capability and enhancement of mechanical properties, while the epoxy functions as a binder that forms an overall solid shape and transfers the stress uniformly onto the fibers inside the matrix. Particularly, epoxies exhibit a low viscosity before curing that helps with fiber wetting and minimizing air voids, aiding in the composite strength. Meanwhile, the low viscosity provides a unique processing versatility such that various options of fabrication techniques can be applied for the industrial manufacture of epoxy composites (see **Table 1**). These characteristics, alongside outstanding chemical and mechanical properties after being cured, let fiber-reinforced epoxies find a broad range of commercial applications in structural materials, commercial and sporting goods, and civil infrastructures, as well as diverse lightweight engineering applications in automotive, marine, aerospace, and wind turbine construction, for which the high strength-to-weight ratio allows for substantial fuel and energy savings [50].

Type	Fiber	Fabrication techniques
Conventional composites	• silica • metal • rubber • basalt • fly ash • aramid • ceramic • natural fiber • carbon black • synthetic fiber • flame retardant fiber • thermoplastic polymer	• extrusion • pultrusion • lamination • solvent casting • filament winding • vacuum bag molding • compression molding • prepreg/autoclave process • vacuum-assisted resin transfer molding
Nanocomposites	• clay • nanofiber • metallic oxide • carbon nanotube • silica nanoparticle • graphene nanoplatelet • boron nitride nanosheet	• melt mixing • shear mixing • electrospinning • solution mixing • ultrasonic mixing • surface modification

Table 1.
Types and fabrication techniques of epoxy-based composites [8, 16, 17, 21, 36, 50–53].

Glass fibers are readily available and relatively cheap on the commodity market, making them the most used fibers in the composite industry. Glass fibers have high tensile strength and excellent insulating properties [54, 55]. Glass fiber-reinforced epoxy composites are often fabricated by distributing discontinuous or continuous glass fibers, randomly oriented, inside the epoxy matrices. By combining the complementary features of glass fiber and epoxy, the ensuing composite provides a superior material for the replacement of heavier metal and ceramic parts in structural and insulation applications. Being another commonly used reinforcement fiber, carbon fiber is stronger, lighter, and more expensive than glass fiber (see **Table 2**). Epoxy adheres well to carbon fiber. The manufacturing of carbon fiber composites usually involves the epoxy impregnation of carbon fabric [34, 36]. The strength of laminated composites depends on the weave pattern, orientation, and weight fraction of the carbon fabric, which is generally stronger than that of the glass fiber reinforcement, leading to more advanced applications.

In addition to synthetic fibers, natural fibers received considerable attention in past decades in view of environmental concerns [56–59]. Natural fibers are mostly sourced from plants or animals and are biodegradable and renewable [60, 61]. Their replacement of synthetic fibers in composite applications can reduce carbon footprint, improve recyclability, and enhance environmental sustainability. Studies have examined epoxy composites reinforced by natural fibers, such as kenaf, coir, sisal, banana, jute, bamboo, cotton, bagasse, flax, ramie, and luffa. These studies suggest these natural fibers reinforce epoxy's physical and mechanical properties and

Material	Bulk density (g/cm^3)	Tensile strength (MPa)	Young's modulus (GPa)	Toughness (mJ/mm^3)
Glass	2.5	1048	62	21
Carbon	1.8	2302	168	46
Aramid	1.4	2273	47	92

Table 2.
Typical properties of commonly used reinforcement fibers [54].

are suitable for low-load applications [62, 63]. Despite growing interest, bio-based materials have seen relatively slow market growth in the consumption of natural fibers [64]. This result provides an opportunity for fostering applications of natural fiber reinforced epoxy composites. However, the development of fiber processing and chemical treatment technologies is needed to improve the performance of epoxy composites with natural fibers.

Given the increasing demand of high-performance materials in lightweight construction and renewable energy generation, significant research efforts are devoted to improving the mechanical properties of fiber-reinforced epoxies and extending their functional applications. Poor damage tolerance and fire resistance are significant obstacles that limit the applications of conventional epoxy-based composites [53]. Notably, their anisotropic nature and weak out-of-plane mechanical properties induce low impact resistance for composites with aligned reinforcement fibers. Meanwhile, the debonding between fiber and matrix under transverse and compression loading can cause delamination that significantly deteriorates the mechanical performances and hence, reduces the service life of composite materials. In addition, the cured epoxy matrix decomposes when exposed to high temperatures (300–400 °C), releasing heat, smoke, and toxic volatiles and byproducts [65]. Managing the fire risk and subsequent environmental hazard is thus crucial for broadening the application range of epoxy composites.

Strategies for composite property improvement focus on developing the matrix, fiber, and the interface of composite materials [66]. Toughening of the epoxy matrix is the most effective way to improve the composite impact strength for structural applications [17, 67]. Standard methodologies for the epoxy toughening include the chemical modification of the epoxy and the integration of toughening agents. In the former scenario, reducing the cross-link density (by increasing the segment length between crosslinks), increasing the molecular weight of epoxy monomers, or modifying the rigid backbone structure to be flexible increases the impact toughness. However, these changes impose significant reductions in other properties, e.g., modulus, tensile and flexural strength, as well as decrease the glass transition temperature that affects the thermal stability. Considering this synergistic effect, the industrial formulations of epoxies often reflect the tradeoff of balancing different properties for end applications. In this regard, blending toughening agents (typically, 5–20 wt%) becomes a more facile approach [16, 17].

Typical toughening agents include liquid rubbers, thermoplastics, hyperbranched polymers, block copolymers, and inorganic nanoparticles, among which the liquid rubber is a more common choice for industrial applications as it is more economical and easier to process [68]. Several types of reactive liquid rubbers, namely, carboxyl-terminated poly (butadiene-co-acrylonitrile), amine-terminated poly (butadiene-co-acrylonitrile), and hydroxyl terminated polybutadiene, have been developed to increase the epoxy toughness [27, 69]. However, these polybutadiene derivatives have unsaturated sites along the backbone that are susceptible to degradation in a high-temperature and oxidative environment. To overcome this deficiency, saturated liquid rubbers, such as polyacrylates, polysiloxane, and polyurethane, are introduced as alternatives that offer better oxidative stability [70]. The liquid rubber toughening is found to be effective for difunctional epoxy resins but ineffective for tri and tetra-functional epoxy resins that have a higher degree of cross-linking, suggesting that the toughening of the matrix increases with increasing the inherent ductility of the epoxy [53, 69].

Thermoplastics constitute another significant type of toughening agents, which enhance the toughness for both difunctional and multifunctional epoxy resins without scarifying other desirable mechanical properties (as is the case for the rubber toughening). Since the early 1980s, several thermoplastics have been applied, including polysulfone, poly (ether imide), poly (ether sulfone), and poly (phenylene oxide) [70, 71]. It is well recognized that the toughening effect of thermoplastic depends on its phase structure inside the matrix. Notably, a co-continuous morphology at a high thermoplastic concentration, in contrast to the discrete morphology, can induce a sharper increase in the fracture toughness with increasing the thermoplastic content [71]. Nevertheless, the manipulation of phase structures (through varying the composition and curing condition) and the optimization of formulations for different applications require additional engineering.

Alongside the development in conventional toughening strategies, block copolymers and nanoparticles emerge as promising additives for epoxy toughening, which gain increasing interest in recent years [17, 19, 72]. Specifically, block copolymers can undergo microphase separation under the thermodynamic driving force. This result imparts a fine-tuning of phase structures at the nanoscale that affects the properties of nanodomains inside the matrix, resulting in stronger enhancement effects compared to conventional agents of larger sizes. Similarly, nanoparticles significantly enlarge the interface between fiber and matrix upon good dispersion inside epoxy. A small amount of nanoparticles can dramatically improve the matrix properties. Moreover, the incorporation of different types of nanoparticles (**Table 1**), and surface modifications [4], lead to epoxy nanocomposites with enhanced electrical conductivity, thermal conductivity, magnetic, flame retardant, and radiation resistant properties [51, 73], expanding their utility in functional applications.

Overall, considering that different types of agents are associated with different toughening mechanisms (functioning at different length scales), a hybridization strategy can be adopted to prepare fiber-reinforced hybrid epoxy composites that enable synergistic toughening [67, 72, 74, 75]. Given the large degree of freedom of the components, the development of multiscale computer simulations [76, 77], machine learning, and experimental studies are needed to expedite the rational design of advanced epoxy composites.

4. Conclusion

Epoxy has high adhesiveness and is widely used in industry. However, no systematic study has been performed on using epoxy as a filler or polymer matrix for composite applications. This chapter has systematically reviewed recent research and advances on epoxy-based composites. Discussion includes analysis of the epoxy type, epoxy function, curing agent, curing process, and performance of epoxy-based composites.

Epoxy resin has a group of reactive polymers that contain epoxy or oxirane groups. There are three major types of epoxy resins: cycloaliphatic, epoxidized, and glycated epoxy resins. Epoxy-based materials are manufactured into the desired shape by injection molding, resin transfer molding, or resin infusion. The desired properties of epoxy resins are generally determined by the combination of an epoxy resin and a curing agent. Three main methods of curing are room-temperature curing, thermal curing, and photocuring. During curing, the chemical reaction of the epoxide groups in the epoxy resin is initiated by a curing agent to form a highly cross-linked network.

As a filler, there are two common approaches to utilize epoxy. One is to functionalize the fiber's surface with epoxy groups so that the fiber's surface can be activated to react with other components in the composites. The other one is to incorporate epoxy as a single domain or phase in the composites with epoxy groups. However, the composites' performance will be influenced by the inherent property of the epoxy and the concentration of the epoxy-included phase.

Epoxy is used as a polymer matrix because of its high specific strength, excellent adhesion, and good dimensional stability. Adding fibers into epoxies improves the thermal and durability properties while reducing the cost for composite applications. However, the addition of fibers can adversely influence the mechanical and physical properties of composites. Toughening the epoxy is an effective way to improve the composite's impact strength for structural applications. Typical toughening agents include thermoplastics, liquid rubbers, block copolymers, and nanoparticles. Liquid rubbers are a more common choice for industrial applications as they are more economical and easier to process.

Challenges and future directions:

1. As a filler, the epoxy's inherent property affects the composites' performance. In addition, the epoxidation process typically involves highly active reagents that can cause uncontrollable side effects that influence the properties of the composites. The selection of appropriate epoxies with a high functionality is a challenge. The epoxy solution concentration, drying process, and advanced epoxy with outstanding mechanical properties can be investigated in the future.

2. As a polymer matrix, the poor interfacial interaction between the epoxy matrix and fibers is a key challenge. Toughening of the epoxy is an effective pathway to address this challenge. The manipulation of phase structures and the optimization of formulations for different applications require additional engineering.

3. Bio-based epoxy can be developed towards the sustainability and recyclability of the composites. Environmentally friendly and economically viable bio-based epoxies are needed.

Acknowledgements

The authors acknowledge the support from the US Department of Energy (DOE) FY 2021 BETO Project under Contract 1.2.1.9 with UT-Battelle LLC, and the US DOE, Office of Energy Efficiency and Renewable Energy, Advanced Manufacturing Office, under CPS Agreement 35714. This book chapter was authored in part by UT-Battelle LLC under contract DE-AC05-00OR22725 with DOE. The US government retains and the publisher, by accepting the article for publication, acknowledges that the US government retains a nonexclusive, paid-up, irrevocable, worldwide license to publish or reproduce the published form of this book chapter, or allow others to do so, for US government purposes. DOE will provide public access to these results of federally sponsored research in accordance with the DOE Public Access Plan (http://energy.gov/downloads/doe-public-access-plan).

Conflict of interest

The authors declare no conflict of interest.

Appendices and nomenclature

BF	Bamboo fiber
BSKP	Bleached softwood kraft pulp
DGEBF	Diglycidyl ether of bisphenol F
ECP	Epoxy chloropropane
ELO	Epoxidized linseed oil
GPTMS	3-glycidoxypropylthrimethoxysilane
PANI	Polyaniline
PBIA	Poly-p-phenylene-benzimidazole-terephthalamide
PLA	Polylactic acid
T-CNC	Tunicate cellulose nanocrystals
TA-ESO	Tannic acid-crosslinked epoxidized soybean oil
UV	Ultraviolet

Author details

Xianhui Zhao[1]*, Shun Lu[2], Wei Li[3], Shuyang Zhang[3], Kai Li[4], Kashif Nawaz[4], Peter Wang[5], Guang Yang[6], Arthur Ragauskas[3,7,8], Soydan Ozcan[5]* and Erin Webb[1]*

1 Environmental Sciences Division, Oak Ridge National Laboratory, Oak Ridge, Tennessee, United States

2 Agricultural and Biosystems Engineering Department, South Dakota State University, Brookings, South Dakota, United States

3 Department of Chemical and Biomolecular Engineering, University of Tennessee, Knoxville, Tennessee, United States

4 Buildings and Transportation Science Division, Oak Ridge National Laboratory, Oak Ridge, Tennessee, United States

5 Manufacturing Science Division, Oak Ridge National Laboratory, Oak Ridge, Tennessee, United States

6 Chemical Sciences Division, Oak Ridge National Laboratory, Oak Ridge, Tennessee, United States

7 Department of Forestry, Wildlife, and Fisheries, Center for Renewable Carbon, The University of Tennessee Institute of Agriculture, Knoxville, Tennessee, United States

8 Joint Institute for Biological Sciences, Biosciences Division, Oak Ridge National Laboratory, Oak Ridge, Tennessee, United States

*Address all correspondence to: zhaox@ornl.gov, ozcans@ornl.gov and webbeg@ornl.gov

References

[1] May C. Epoxy Resins: Chemistry and Technology. Boca Raton, Florida, USA: Routledge; 2018

[2] Wang Y, Chen S, Chen X, Lu Y, Miao M, Zhang D. Controllability of epoxy equivalent weight and performance of hyperbranched epoxy resins. Composites Part B: Engineering. 2019;**160**:615-625

[3] Jin F-L, Li X, Park S-J. Synthesis and application of epoxy resins: A review. Journal of Industrial and Engineering Chemistry. 2015;**29**:1-11

[4] Gu H, Ma C, Gu J, Guo J, Yan X, Huang J, et al. An overview of multifunctional epoxy nanocomposites. Journal of Materials Chemistry C. 2016;**4**:5890-5906

[5] Ahmadi Z. Epoxy in nanotechnology: A short review. Progress in Organic Coatings. 2019;**132**:445-448

[6] Song B, Wang T, Wang L, Liu H, Mai X, Wang X, et al. Interfacially reinforced carbon fiber/epoxy composite laminates via in-situ synthesized graphitic carbon nitride (g-C3N4). Composites Part B: Engineering. 2019;**158**:259-268

[7] Huang J-K, Young W-B. The mechanical, hygral, and interfacial strength of continuous bamboo fiber reinforced epoxy composites. Composites Part B: Engineering. 2019;**166**:272-283

[8] Chen Y, Zhang H-B, Yang Y, Wang M, Cao A, Yu Z-Z. High-performance epoxy nanocomposites reinforced with three-dimensional carbon nanotube sponge for electromagnetic interference shielding. Advanced Functional Materials. 2016;**26**:447-455

[9] Hu Z, Zhang D, Lu F, Yuan W, Xu X, Zhang Q, et al. Multistimuli-responsive intrinsic self-healing epoxy resin constructed by host–guest interactions. Macromolecules. 2018;**51**:5294-5303

[10] Miao J-T, Yuan L, Guan Q, Liang G, Gu A. Biobased heat resistant epoxy resin with extremely high biomass content from 2, 5-furandicarboxylic acid and eugenol. ACS Sustainable Chemistry & Engineering. 2017;**5**:7003-7011

[11] Zhao X, Li K, Wang Y, Tekinalp H, Larsen G, Rasmussen D, et al. High-strength polylactic acid (PLA) biocomposites reinforced by epoxy-modified pine fibers. ACS Sustainable Chemistry & Engineering. 2020;**8**:13236-13247

[12] Sujaritjun W, Uawongsuwan P, Pivsa-Art W, Hamada H. Mechanical property of surface modified natural fiber reinforced PLA biocomposites. Energy Procedia. 2013;**34**:664-672

[13] Kyutoku H, Maeda N, Sakamoto H, Nishimura H, Yamada K. Effect of surface treatment of cellulose fiber (CF) on durability of PLA/CF bio-composites. Carbohydrate Polymers. 2019;**203**:95-102

[14] Li K, McGrady D, Zhao X, Ker D, Tekinalp H, He X, et al. Surface-modified and oven-dried microfibrillated cellulose reinforced biocomposites: Cellulose network enabled high performance. Carbohydrate Polymers. 2021;**256**:117525

[15] Vidil T, Tournilhac F, Musso S, Robisson A, Leibler L. Control of reactions and network structures of epoxy thermosets. Progress in Polymer Science. 2016;**62**:126-179

[16] Paluvai NR, Mohanty S, Nayak S. Synthesis and modifications of epoxy

resins and their composites: A review. Polymer-Plastics Technology and Engineering. 2014;**53**:1723-1758

[17] Domun N, Hadavinia H, Zhang T, Sainsbury T, Liaghat G, Vahid S. Improving the fracture toughness and the strength of epoxy using nanomaterials–a review of the current status. Nanoscale. 2015;**7**:10294-10329

[18] Kim BC, Park SW. Fracture toughness of the nano-particle reinforced epoxy composite. Composite Structures. 2008;**86**:69-77

[19] Huang F, Liu Y, Zhang X, Wei G, Gao J, Song Z, et al. Effect of elastomeric nanoparticles on toughness and heat resistance of epoxy resins. Macromolecular Rapid Communications. 2002;**23**:786-790

[20] Garg AC, Mai Y-W. Failure mechanisms in toughened epoxy resins—A review. Composites Science and Technology. 1988;**31**:179-223

[21] Nouri-Borujerdi A, Kazemi-Ranjbar S. Thermal and electrical conductivity of a graphene-based hybrid filler epoxy composite. Journal of Materials Science. 2021;**56**:15151-15161

[22] Baroncini EA, Kumar Yadav S, Palmese GR, J.F. Stanzione III, Recent advances in bio-based epoxy resins and bio-based epoxy curing agents. Journal of Applied Polymer Science. 2016;**133**(45):44103

[23] Shi Q, Yu K, Kuang X, Mu X, Dunn CK, Dunn ML, et al. Recyclable 3D printing of vitrimer epoxy. Materials Horizons. 2017;**4**:598-607

[24] Zhao X, Korey M, Li K, Copenhaver K, Tekinalp H, Celik S, et al. Plastic waste upcycling toward a circular economy. Chemical Engineering Journal. 2022;**428**:131928

[25] Kumar A, Anant R, Kumar K, Chauhan SS, Kumar S, Kumar R. Anticorrosive and electromagnetic shielding response of a graphene/TiO_2–epoxy nanocomposite with enhanced mechanical properties. RSC Advances. 2016;**6**:113405-113414

[26] Pulikkalparambil H, Siengchin S, Parameswaranpillai J. Corrosion protective self-healing epoxy resin coatings based on inhibitor and polymeric healing agents encapsulated in organic and inorganic micro and nanocontainers. Nano-Structures & Nano-Objects. 2018;**16**:381-395

[27] Chikhi N, Fellahi S, Bakar M. Modification of epoxy resin using reactive liquid (ATBN) rubber. European Polymer Journal. 2002;**38**:251-264

[28] Griffin S, Askarinejad H, Farrant B. Evaluation of epoxy injection method for concrete crack repair. International Journal of Structural and Civil Engneering Research. 2017;**6**:177-181

[29] Yousef S, Eimontas J, Subadra SP, Striūgas N. Functionalization of char derived from pyrolysis of metallised food packaging plastics waste and its application as a filler in fiberglass/epoxy composites. Process Safety and Environmental Protection. 2021;**147**:723-733

[30] Lee H, Neville K. Handbook of Epoxy Resins. New York, New York, USA: McGraw-Hill; 1967

[31] Peerzada M, Abbasi S, Lau KT, Hameed N. Additive manufacturing of epoxy resins: Materials, methods, and latest trends. Industrial & Engineering Chemistry Research. 2020;**59**:6375-6390

[32] Dagdag O, El Harfi A, Essamri A, El Gouri M, Chraibi S, Assouag M, et al. Phosphorous-based epoxy resin

composition as an effective anticorrosive coating for steel. International Journal of Industrial Chemistry. 2018;**9**:231-240

[33] Calza P, Rigo L, Sangermano M. Investigations of photocatalytic activities of photosensitive semiconductors dispersed into epoxy matrix. Applied Catalysis B: Environmental. 2011;**106**:657-663

[34] Zhang K, Gu Y, Zhang Z. Effect of rapid curing process on the properties of carbon fiber/epoxy composite fabricated using vacuum assisted resin infusion molding. Materials & Design (1980-2015). 2014;**54**:624-631

[35] Zhan J, Li J, Wang G, Guan Y, Zhao G, Lin J, et al. Review on the performances, foaming and injection molding simulation of natural fiber composites. Polymer Composites. 2021;**42**:1305-1324

[36] Cheng QF, Wang JP, Wen JJ, Liu CH, Jiang KL, Li QQ, et al. Carbon nanotube/epoxy composites fabricated by resin transfer molding. Carbon. 2010;**48**:260-266

[37] Mezeix L, Seman A, Nasir MNM, Aminanda Y, Rivai A, Castanié B, et al. Spring-back simulation of unidirectional carbon/epoxy flat laminate composite manufactured through autoclave process. Composite Structures. 2015;**124**:196-205

[38] Ma S, Liu X, Fan L, Jiang Y, Cao L, Tang Z, et al. Synthesis and properties of a bio-based epoxy resin with high epoxy value and low viscosity. ChemSusChem. 2014;**7**:555-562

[39] Fang Q, Yao J, Niu K, Tang J, Wei Y, Guo Q, et al. Effect of molecular weight of self-emulsifying amphiphilic epoxy sizing emulsions on the carbon fibres and interfacial properties of their composites. Polymers. 2020;**12**:2439

[40] Iijima T, Yoshioka N, Tomoi M. Effect of cross-link density on modification of epoxy resins with reactive acrylic elastomers. European Polymer Journal. 1992;**28**:573-581

[41] Ozeren Ozgul E, Ozkul MH. Effects of epoxy, hardener, and diluent types on the hardened state properties of epoxy mortars. Construction and Building Materials. 2018;**187**:360-370

[42] Kuang X, Zhao Z, Chen K, Fang D, Kang G, Qi HJ. High-speed 3D printing of high-performance thermosetting polymers via two-stage curing. Macromolecular Rapid Communications. 2018;**39**:1700809

[43] Griffini G, Invernizzi M, Levi M, Natale G, Postiglione G, Turri S. 3D-printable CFR polymer composites with dual-cure sequential IPNs. Polymer. 2016;**91**:174-179

[44] Cheng Z, Li B, Huang J, Chen T, Liu Y, Wang X, et al. Covalent modification of aramid fibers' surface via direct fluorination to enhance composite interfacial properties. Materials & Design. 2016;**106**:216-225

[45] Fei G, Wang Y, Wang H, Ma Y, Guo Q, Huang W, et al. Fabrication of bacterial cellulose/polyaniline nanocomposite paper with excellent conductivity, strength, and flexibility. ACS Sustainable Chemistry & Engineering. 2019;7:8215-8225

[46] Cao L, Gong Z, Xu C, Chen Y. Mechanical strong and recyclable rubber nanocomposites with sustainable cellulose nanocrystals and interfacial exchangeable bonds. ACS Sustainable Chemistry & Engineering. 2021;**9**:9409-9417

[47] Immonen K, Anttila U, Wikström L. Coupling of PLA and bleached softwood Kraft pulp (BSKP) for enhanced properties of biocomposites. Journal of

Thermoplastic Composite Materials. 2019;**32**:328-341

[48] Liu W, Qiu J, Chen T, Fei M, Qiu R, Sakai E. Regulating tannic acid-crosslinked epoxidized soybean oil oligomers for strengthening and toughening bamboo fibers-reinforced poly (lactic acid) biocomposites. Composites Science and Technology. 2019;**181**:107709

[49] Karak N. Overview of Epoxies and their Thermosets, Sustainable Epoxy Thermosets and Nanocomposites. Washington, D.C., USA: American Chemical Society; 2021. pp. 1-36

[50] Rajak DK, Pagar DD, Menezes PL, Linul E. Fiber-reinforced polymer composites: Manufacturing, properties, and applications. Polymers. 2019;**11**: 1667

[51] Li R, Gu Y, Zhang G, Yang Z, Li M, Zhang Z. Radiation shielding property of structural polymer composite: Continuous basalt fiber reinforced epoxy matrix composite containing erbium oxide. Composites Science and Technology. 2017;**143**:67-74

[52] Kausar A, Anwar Z, Muhammad B. Recent developments in epoxy/graphite, epoxy/graphene, and epoxy/graphene nanoplatelet composites: A comparative review. Polymer-Plastics Technology and Engineering. 2016;**55**:1192-1210

[53] Ratna D. Epoxy Composites: Impact Resistance and Flame Retardancy. Shawbury, Shrewsbury, Shropshire, UK: iSmithers Rapra Publishing; 2007

[54] Yao Y, Zhu D, Zhang H, Li G, Mobasher B. Tensile behaviors of basalt, carbon, glass, and aramid fabrics under various strain rates. Journal of Materials in Civil Engineering. 2016;**28**:04016081

[55] Ogbonna VE, Popoola API, Popoola OM, Adeosun SO. A review on corrosion, mechanical, and electrical properties of glass fiber-reinforced epoxy composites for high-voltage insulator core rod applications: Challenges and recommendations. Polymer Bulletin. 2021:1-28. DOI: 10.1007/s00289-021-03846-z

[56] Wang L, Gardner DJ, Wang J, Yang Y, Tekinalp HL, Tajvidi M, et al. Towards industrial-scale production of cellulose nanocomposites using melt processing: A critical review on structure-processing-property relationships. Composites Part B: Engineering. 2020;**201**:108297

[57] Bhagia S, Bornani K, Agrawal R, Satlewal A, Ďurkovič J, Lagaňa R, et al. Critical review of FDM 3D printing of PLA biocomposites filled with biomass resources, characterization, biodegradability, upcycling and opportunities for biorefineries. Applied Materials Today. 2021;**24**:101078

[58] Zhao X, Copenhaver K, Wang L, Korey M, Gardner DJ, Li K, et al. Recycling of natural fiber composites: Challenges and opportunities. Resources, Conservation and Recycling. 2022;**177**:105962

[59] Zhao X, Li K, Wang Y, Tekinalp H, Richard A, Webb E, et al. Bio-treatment of poplar via amino acid for interface control in biocomposites. Composites Part B: Engineering. 2020;**199**:108276

[60] Meng X, Scheidemantle B, Li M, Wang YY, Zhao X, Toro-Gonzalez M, et al. Synthesis, characterization, and utilization of a lignin-based adsorbent for effective removal of azo dye from aqueous solution. ACS Omega. 2020;**5**:2865-2877

[61] Zhao X, Tekinalp H, Meng X, Ker D, Benson B, Pu Y, et al. Poplar as biofiber reinforcement in composites for large-scale 3D printing. ACS Applied Bio Materials. 2019;**2**:4557-4570

[62] Mohammed L, Ansari MNM, Pua G, Jawaid M, Islam MS. A review on natural Fiber reinforced polymer composite and its applications. International Journal of Polymer Science. 2015;**2015**:243947

[63] Jeyapragash R, Srinivasan V, Sathiyamurthy S. Mechanical properties of natural fiber/particulate reinforced epoxy composites – A review of the literature. Materials Today: Proceedings. 2020;**22**:1223-1227

[64] Wang L, Li K, Copenhaver K, Mackay S, Lamm ME, Zhao X, et al. Review on nonconventional fibrillation methods of producing cellulose Nanofibrils and their applications. Biomacromolecules. 2021;**22**:4037-4059

[65] Mouritz AP, Gibson AG. Fire Properties of Polymer Composite Materials. Dordrecht, The Netherlands: Springer Science and Business Media; 2007

[66] Copenhaver K, Li K, Lamm ME, Walker C, Johnson D, Han Y, et al. Recycled cardboard containers as a low energy source for cellulose Nanofibrils and their use in poly(l-lactide) nanocomposites. ACS Sustainable Chemistry & Engineering. 2021;**9**:13460-13470

[67] Carolan D, Ivankovic A, Kinloch A, Sprenger S, Taylor A. Toughening of epoxy-based hybrid nanocomposites. Polymer. 2016;**97**:179-190

[68] Wang Y, Li K, Zhao X, Tekinalp H, Li T, Ozcan S. Toughening by nanodroplets: Polymer–droplet biocomposite with anomalous toughness. Macromolecules. 2020;**53**:4568-4576

[69] Ratna D, Banthia AK. Rubber toughened epoxy. Macromolecular Research. 2004;**12**:11-21

[70] Cho J, Hwang J, Cho K, An J, Park C. Effects of morphology on toughening of tetrafunctional epoxy resins with poly (ether imide). Polymer. 1993;**34**:4832-4836

[71] Ma H, Aravand MA, Falzon BG. Phase morphology and mechanical properties of polyetherimide modified epoxy resins: A comparative study. Polymer. 2019;**179**:121640

[72] Klingler A, Bajpai A, Wetzel B. The effect of block copolymer and core-shell rubber hybrid toughening on morphology and fracture of epoxy-based fibre reinforced composites. Engineering Fracture Mechanics. 2018;**203**:81-101

[73] Zhang J, Kong Q, Wang D-Y. Simultaneously improving the fire safety and mechanical properties of epoxy resin with Fe-CNTs via large-scale preparation. Journal of Materials Chemistry A. 2018;**6**:6376-6386

[74] Wang J, Xue Z, Li Y, Li G, Wang Y, Zhong W-H, et al. Synergistically effects of copolymer and core-shell particles for toughening epoxy. Polymer. 2018;**140**:39-46

[75] Li K, Wang Y, Rowe M, Zhao X, Li T, Tekinalp H, et al. Poly(lactic acid) toughening through chain end engineering. ACS Applied Polymer Materials. 2020;**2**:411-417

[76] Li W, Jana PK, Behbahani AF, Kritikos G, Schneider L, Polińska P, et al. Dynamics of long entangled Polyisoprene melts via multiscale modeling. Macromolecules. 2021;**54**:8693-8713

[77] Wu X, Aramoon A, El-Awady JA. Hierarchical multiscale approach for modeling the deformation and failure of epoxy-based polymer matrix composites. The Journal of Physical Chemistry B. 2020;**124**:11928-11938

Section 2

Research on Epoxy Based Composites

Chapter 2

A Review on Mechanical Properties of Epoxy-Glass Composites Reinforced with Nanoclay

Shanti Kiran Zhade, Syam Kumar Chokka, V. Suresh Babu and K.V. Sai Srinadh

Abstract

Polymer nanocomposites are currently one of the most rapidly growing families of materials, and they are finding use in a wide range of industrial applications, including aerospace and defense. The broad usage of composites is because of their consolidated mechanical properties. Glass fiber reinforced epoxy composites are available for the last few decades. The idea of adding nano clay into it has emerged in the late first decade of this century. This study is aimed at reporting the effects of the addition of nano clay into GFRP on its mechanical properties. The new composite formed is epoxy-glass composites reinforced with nano clay (EGCN). Nano clay has a crystal structure that facilitates the formation of intercalated and exfoliated mixture with liquid epoxy during mixing which results in good dispersion of Nano clay thereby resulting in improved mechanical properties compared to GFRP. The work done by several researchers in this area and the results obtained are reported in this article. The improved mechanisms of failures were discussed with the addition of nano clays.

Keywords: nano clay particles, GFRP composites, mechanical properties, interlaminar shear strength

1. Introduction

In the current era, much effort is invested in developing new composite materials which are superior to existing materials in terms of their mechanical and physical properties. Quite a large number of studies have been published in the area of behavior, characterization, and modeling of composite materials, from metal matrix composites [1–3] to polymer matrix composites [4–7]. There are several definitions for composite materials, but the common feature of each definition is the presence of two or more constituents with an interface between them. Traditional metal matrix-based composites are made of heavy materials. In the aerospace and automobile sector, the fuel consumption is proportional to the weight of the body of the vehicle. A study by A. K. Dhingra [8] has shown that a minimum of 20% of the cost is saved if polymer composites (PMCs) replace the metal structures and the operating and maintenance costs are also very low. Chakraverty AP [9] stated that Polymer composites are easy to repair, have good durability, and maintenance is simple. There is a consistent requirement for composites in the industries with the invention of new applications. Glass fiber-epoxy composites are widely used in the making of aircraft and automobile body parts and are not only limited to these fields but also used in ship building, structural applications in civil engineering, pipes for the transport of liquids, electrical insulators in reactors, etc. GFRPs have been in use since 1936. As the requirements for weight reduction continuously increase the mechanical properties of GFRPs can be tailored by adding micro fillers and it further evolved by adding nanofillers. The epoxy resin in GFRP firmly holds the fibers together and helps in uniform load distribution throughout the composite.

Nanocomposites are materials that are created by introducing nanoparticles into a matrix. There is a drastic improvement in mechanical properties with the addition of nanomaterials into various matrix materials. In general, the content of nanoparticles that can be added to the composite ranges between 0.5% and 5%. It is because of the high surface area of nanomaterials at a given weight content compared to the micron-sized powder of the same material. Plenty of research is in progress to develop nanocomposites with multiple functionalities. The term "polymer nanocomposite" broadly describes any number of multicomponent systems where the primary component is the polymer matrix and the filler material has at least one dimension below 100 nm [10]. Polymer nanocomposites are generally lightweight, require low filler loading, are often easy to process, and provide property enhancements extending orders of magnitude beyond those realized with traditional composites.

Filler is a term that encompasses a vast number of materials and plays a significant role in improving composite properties. Fillers help minimize cost, enhance properties, and improve the composites. Fillers also increase the mechanical properties and reduce shrinkage of the composites during curing. Proper selection of matrix and filler combination will lead to the creation of composites with high mechanical and thermo-mechanical properties which are comparable to metals.

Montmorillonite is natural clay with a high charge density. Charge density is the total number of cations in between the silicate layers of montmorillonite which can be substituted with organic cations. Montmorillonite nanoparticles are naturally hydrophilic but if treated with alkylammonium ions, the particles become organophilic. The organically treated montmorillonite when dispersed in liquids like epoxy forms gels [11, 12]. The length of the ammonium ions has a strong impact on the resulting structure of nanocomposites. Lan et al. [13] showed that alkylammonium ions with chain lengths larger than eight carbon atoms favor the synthesis of exfoliated nanocomposites, whereas alkylammonium ions with shorter chains led to the formation of intercalated nanocomposites. Alkylammonium ions based on secondary amines have also been successfully used [14]. A schematic diagram showing the substitution of alkylammonium ions in place of interlayer cations is shown in **Figure 1**. The structure of the organic cations between silicate layers depends on the charge density of clay [15]. In **Figure 1**. alkylammonium ions adopt a paraffin type of structure due to which the spacing between the clay layers increased by about 10 A°. Alkylammonium ions permit lowering the surface energy of clay so that organic species with different polarities can get intercalated between the clay layers.

Based on the above discussion, it is observed that out of the three types of surface modifiers alkyl ammonium ions are the most popular because they have a higher affinity with silicate layers compared to amino acids and silanes. Depending on the layer charge density of the clay, alkyl ammonium ions may adopt different structures

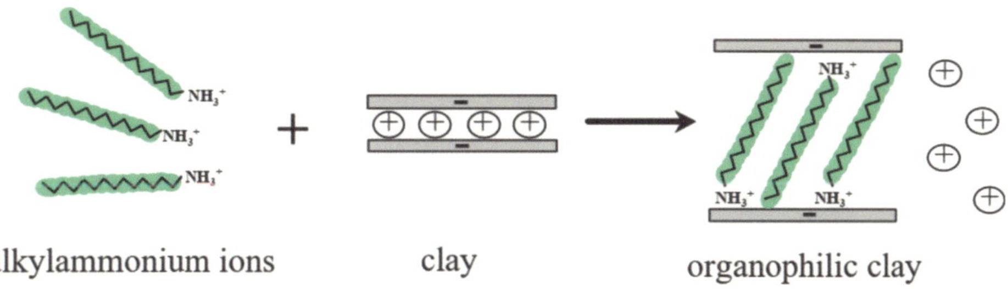

Figure 1.
The cation-exchange process between alkylammonium ions and cations initially intercalated between the clay layers.

between the clay layers. Alkyl ammonium ions reduce the electrostatic interactions between silicate layers thus facilitating the diffusion of a polymer molecule between clay platelets or galleries [16].

2. Mechanical properties

2.1 Tensile properties

EGCN exhibited 54% improvement in modulus at 10 wt% addition of octadecyl ammonium treated fluorohectorite (ME-ODA) but there was a 36% decrease in strength while ductility was also reduced (shown **Figure 2a**). The stress–strain curve of GRE exhibited ductile behavior, and the EGCN exhibited brittle behavior [17]. Bozkurt et al. [18] reported that when MMT is added to epoxy-noncrimp glass fabric composite, up to 6 wt% there was no improvement in strength and stiffness while both decreased beyond 6 wt%. This unchanging behavior is attributed to the dominant effect of noncrimp glass fibers over the nanoclay effect (shown **Figure 2b**). Shi et al. [19] reported the effect of "Magnetic stirring" and the high shear mixing technique (HSMT) on the tensile behavior of EGCN. When the magnetic stirring method was used, the increment in modulus of EGCN was about 19.4%, 22.2%, and 27.7% at 1, 2, and 3 wt% of nanoclay respectively. The increased modulus is credited to the good dispersion of clay layers. At 1 wt% nanoclay, the tensile properties were compared between composites; one composite consists of an epoxy-clay mixture processed by magnetic stirring, and the other by mechanical stirring. There was about a 7.9% and 5.7% increase in σ**UTS** and modulus for the EGCN with matrix processed by mechanical stirring (shown **Figure 2c**). The epoxy molecular chains were prevented from moving when the load was applied. The clay layers hindered the molecular chains because of strong adhesion and chemical bond between organoions and epoxy, thereby enhancing the stiffness of the laminate. This mechanism of clay particles hindering epoxy molecular movement was also described by other researchers [20–22]. The formation of clay aggregates at low clay contents, i.e., at 2 and 3 wt% clay addition was also reported in some literature [23–25]. The increased tensile properties with the addition of various surface-modified nanoclays under different mixing conditions and composite making methods are given in **Table 1**.

Voids are formed while mixing nanoclay and hardener, and increase with clay content; due to an increase in the viscosity of the mixture, the removal of these gas bubbles becomes difficult when kept in degassing chamber. In addition to aggregates and voids, there is a possibility of a decrease in strength by other means, that is,

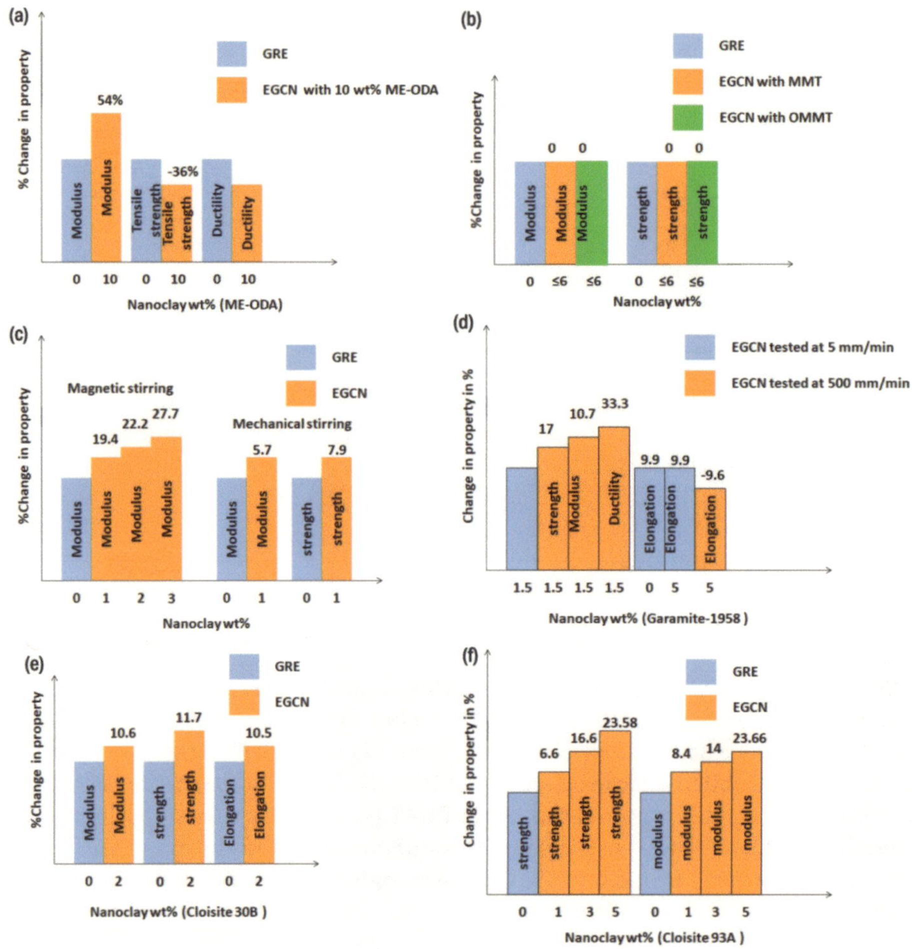

Figure 2.
Changes in tensile properties of EGCN'S at various conditions.

through interruption of crosslinking of chains by silicate layers as a result of the reaction of epoxy molecules with organoions, which breaks the continuity of the crosslinks, this claim is not yet fully established though [35, 36]. The laminate fabricated from the matrix which was prepared by HSMT provided the enhancement in strength and modulus by 7.9% and 5.7% as compared to the laminates made by using the matrix prepared by direct mixing technique (DMT) [19].

Gurusideswar and Velmurugan [29] carried out tension tests on laminates with the addition of Garamite-1958 (alkyl ammonium treated clay) at crosshead speeds of 0.5, 5, 50, and 500 mm/min. The stress–strain plot for EGCN at 1.5 wt% of nanoclay and a testing speed of 5 mm/min was linear elastic with 9.9% elongation and failed suddenly. At 500 mm/min there was a rise in strength, modulus, and ductility by about 17%, 10.7%, and 33.3% compared to the values at quasi-static loading, i.e., at 5 mm/min (shown **Figure 2d**). σ**UTS** is more sensitive to strain rate compared to the modulus which is due to the dominant behavior of fibers in strength, whereas the modulus is influenced by clay. The same behavior was exhibited by glass/epoxy composite (GRE). The increase in clay content by up to 5 wt% did not change the elongation (i.e., 9.9%) at quasi-static loading, but at high strain rates.

Author, Year	Nanoclay	Nanoclay wt%	Glass Fiber	Observations
Haque et al. [26]	Nanomer I.28E	0, 1, 2, 5, 10 wt%	—	At 1 wt% nanosilicates, there was 44, 24, and 23% improvement in ILSS, σ_f, and fracture toughness compared to GRE.
Bozkurt et al. [18]	MMT	0, 1, 3, 6, and 10%wt	40 – 44 vol %	i. Up to 6 wt% nanoclay addition, σ_{UTS} and modulus were not changed. ii. At 6 wt% of nanoclay, σ_f and modulus were improved by 16% and 13%. iii. At 10 wt% of nanoclay, fracture toughness was improved by 5%
Shi et al. [19]	Cloisite 30B	0, 1, 2 and 3 wt%	55 – 58 vol%	**At 1 wt%** i. The tensile, flexural, and compressive modulus increased by 21%, 27%, 15% respectively. ii. The tensile, flexural, and compressive strength were increased by 18%, 25%, and 30% respectively. iii. ILSS was increased by 25% and Impact strength was increased by 6%.
Zulfli and Chow [27]	Nanomer 1.28E	0, 2, 4, 6, 8 wt%	4 layers	i. At 4 wt% of clay, σ_f and modulus were improved by 19% and 9%. ii. At 2 wt% of nanoclay, fracture toughness was improved by 111%. iii. At 4 wt% of nanoc\lay, impact strength was improved by 46%
Kanny and Mohan [28]	Cloisite 30B	0, 1, 2, 3, 4, 5 wt%	6 layers	i. At 3 wt.% of nanoclay, there was about a 9%, 21%, and 15% increase in tensile strength, modulus, and elongation.
Gurusideswar and Velmurugan [29]	Garamite_ 1958	1.5, 3, and 5 wt%	—	i. At 1.5% clay, the modulus and strength increased by 5%, 3% compared to neat epoxy. ii. The modulus and strength of 1.5 wt% EGCN with 1.5 wt% clay 5 mm/min crosshead speed increased by 8%, 1% compared to GRE.
Kornmann et al. [17]	ME-100	10 wt%	55 vol%	At 10 wt% of nanoclay. i. Flexural modulus increased by 8% ii. σ_f decreased by 27% iii. Flexural strain increased by 19%.
Sharma et al. [30]	Cloisite 30B	1, 3, and 5 wt%	—	i. Up to 3 wt% of clay, σ_{UTS} was increased and decreased beyond that. ii. σ_f was increased up to 5 wt% of nanoclay addition.
Manfredi et al. [12]	i) Cloisite 30B ii) Cloisite 10A	3, 5 wt%	30 vol%	a. At 30 vol% glass fiber, and 5 wt% of Cloisite 10A i. Flexural modulus and strength were increased by 20% and 29%. ii. ILSS was increased by 8% iii. The impact strength was improved by 23% b. The properties were not improved by the Cloisite 30B

Author, Year	Nanoclay	Nanoclay wt%	Glass Fiber	Observations
Gurusideswar and Velmurugan [31]	Garamite_1958	0, 1.5, 3 wt%	—	a. At quasi static strain rate of 0.00167 s^{-1} with corresponding loading speed of 5 mm/min. i. 15% improvement in young's modulus was observed at 3 wt% of nanoclay ii. 9% improvement in σ_{UTS} at 1.5 wt% of nanoclay. b. At the strain rate of 445 s^{-1} and 1.5 wt% of nanoclay the strength was improved by 84% c. At 0 wt% clay and i. At 0.00167 s^{-1} strain rate the strength and modulus were 314.92 MPa, 18.09 GPa ii. At 315 s^{-1} the values were improved by 34%, 58%. iii. At 385 s^{-1} strain rate the values improved by 51%, 92%. iv. At 445 s^{-1} strain rate the values were improved by 67%, 106%.
Krushnamurty et al. [32]	Nanomer I.30E	0, 3 wt%	40, 50, 60, 70	At 3 wt% of addition of nanoclay, the σ_{UTS} was improved by i. 21% at 40% of fiber volume ii. 13% at 50% of fiber volume iii. 7% at 60% of fiber volume iv. –2% at 70% of fiber volume At 3 wt % addition of nanoclay, the σ_f was improved by i. 20% at 40% of fiber volume ii. 14% at 50% of fiber volume iii. 8% at 60% of fiber volume iv. 3% at 70% of fiber volume
Jeyakumar et al. [33]	Cloisite 93	1, 3, 5, 7 wt%	—	i. Maximum improvement in σ_{UTS} is 30% at 5 wt%. ii. Maximum improvement in tensile modulus is 32% at 5 wt%. iii. Maximum improvement in σ_f is 50% at 5 wt%. iv. The maximum improvement in flexural modulus is 116% at 5 wt%. v. The maximum improvement in impact strength is 42% at 3 wt%. vi. The maximum improvement in fracture toughness is 136% at 5 wt%.

Author, Year	Nanoclay	Nanoclay wt%	Glass Fiber	Observations
Achutha et al. [34]	OMMT	0, 2, 4 wt%	40, 50, 60 wt%	The maximum improvement in the property of EGCN with the addition of 4 wt% of OMMT and 60 wt% glass fiber compared to GRE with 60 wt% fiber i. σ_{UTS} improved by 11.5% at room temp ii. σ_f improved by 4.5% at room temp iii. For specimens soaked in cold water for 70 days, there was a 7.5% decrease in tensile strength iv. For specimens soaked in cold water for 70 days, there was a 10% decrease in σ_f v. For specimens soaked in boiling water for 2 hrs, there was a 9% decrease in tensile strength vi. For specimens soaked in boiling water for 2 hrs, there was a 12.5% decrease in σ_f
Prabhakar et al. [11]	a. MMT b. Nanomer I.28E	3 wt%	40 wt%	a. Improvement in tensile, flexural, and impact strength values of EGCN compared to GRE i. With the addition of MMT are −54.4%, −19.2%, −20.7% ii. With the addition of OMMT are −11.5%, −33.8%, −20.7% iii. With the addition of MMT and silane treated glass fiber are −25.36%, −9.2%, −59% iv. With the addition of OMMT and silane treated glass fiber are 6%, −9.9%, −1.8% v. With the addition of MMT and acid treated glass fiber are −30.7%, −28%, 2.2% With the addition of OMMT and acid treated glass fiber are −43%, −58.5%, −59.2% i. Improvement in tensile, flexural, and impact strength values of GRE with addition of silane treated glass fiber −16.6%, 12.5%, 33.75%. ii. Improvement in tensile, flexural, and impact strength values of GRE with the addition of acid treated glass fiber −26.5%, −5.3%, −40.3%

Table 1.
Effect of Nano clay on various properties.

the elongation has reduced (i.e. 9.6% at 500 mm/min). This is attributed to the high brittleness induced at high clay addition. At 1.5 wt% of nanoclay addition, the inversely proportional behavior between elongation and strain rate was not observed, unlike the case of 5 wt% clay added composite. The increase in strength is mainly attributed to the presence of fibers and the increase in modulus is mainly attributed to the silicate platelets which restrict the movement of epoxy molecules [37–41].

The optimum value of clay content is 1.5 wt%, whereas all tensile characteristics were improved at a high strain rate, the slight decrease in properties above 1.5 wt% nanoclay is attributed to agglomeration and a weak interfacial bond between epoxy and clay. An increase in strain rate in the range of 0.0006 s^{-1}-0.6 s^{-1} increased the strength and elongation of GRE. Okoli and Smith [42] reported that there was a decrease in percentage elongation when GRE specimens were tested at various strain rates. Okoli and Smith [42] added that the decrease in elongation at high strain rates is explained with the help of Eyring theory of viscosity; while formulating this theory, an assumption has been made which states that the molecules of polymer need to cross the potential energy barriers to deform when a load is applied. Based on this assumption a linear model is developed which states that the plot between yield stress and a logarithm of strain rate is linear. This increase in yield stress with the logarithm of strain rate implies decreased plastic deformation of the matrix due to decreased movement of crosslinked epoxy molecules at high strain rates. The constrained movement of molecules is ascribed to the lack of time available for the molecules to relax at high strain rates [43–45]. But according to Gurusideswar and velmurugan [29], this effect was absent at 500 mm/min as there was an increased elongation for GRE at 500 mm/min compared to the elongation at quasi-static loading rate, which implies that 500 mm/min is not high enough to restrict the molecules' relaxation. Hussain F [45] stated that the high modulus of clay is also one of the attributes for an increase in tensile properties and improved deformation mechanisms. Li X et al. [46] Reported that the exfoliated structures have a high surface area of contact between silicate platelets and resin; therefore the transfer of load to clay platelets also will be more compared to the load transferred in intercalated structures. Withers et al. [47] reported an 11.7%, 10.6%, and 10.5% increase in strength, modulus, and elongation with 2 wt% of Cloisite 30B loading into glass-epoxy due to exfoliated morphology (shown **Figure 2e**).

Gurusideswar and Velmurugan [31] reported the behavior of EGCN with the addition of GARMITE-1958 and testing speeds varying between quasi-static rate of 0.00167 s^{-1} to very high strain rates of 315 s^{-1}, 385 s^{-1}, 445 s^{-1} which are far higher compared to the strain rates in the range of 0.0001-0.1 s^{-1}. GRE exhibited about 106% and 67% improvement in modulus and strength at 445 s^{-1} compared to quasi-static conditions. EGCN exhibited about a 150% rise in modulus and 84% rise in strength at 1.5 wt% clay addition and 445 s^{-1} strain rate. This substantial rise in modulus and strength of EGCN is attributed to viscoelastic nature, damage accumulation behavior of epoxy which was also reported by Brown et al. [48] for GRE, restriction of polymers chain mobility in the matrix and at the fiber-matrix interface due to good adhesion between clay platelets and epoxy allowed better stress transfer to all the fibers. Similar findings were reported by many authors [49, 50]. Jeyakumar et al. [33] reported the mechanical properties of EGCNs with the addition of Cloisite 93A into epoxy-glass. Nanoclay was mixed into acetone using a mechanical stirrer for 30 min. Epoxy resin of the required weight was added to the acetone-clay mixture heated to 80°C and mixed for 1 hr. During this process, acetone gets evaporated and the epoxy clay mixture remains. The remaining mixture is ultrasonicated for uniform mixing.

The testing results of the prepared samples showed that the σ**UTS** improved by 6.6%, 16.6%, and 23.58% at 1, 3, and 5 wt% of nanoclay, whereas tensile modulus improved by 8.4%, 14%, and 23.66%. With the further addition of nanoclay, the decreasing trend started (shown in **Figure 2f**).

Achutha et al. [34] attempted to optimize the parameters such as nanoclay wt% and glass fiber content in EGCN. In addition, the samples were also subjected to hydrothermal conditions. A set of samples were soaked in cold water for 70 days and dried and another set of samples was boiled in hot water for 2 hrs and dried. It was reported from these studies that hydrothermal aging conditions showed 42.69% contribution to tensile properties whereas nanoclay content showed 24.57% contribution and fiber content showed 30.23% contribution. Achutha et al. [34] reported that nanoclay does not act as a load-bearing instrument but it warrants load transfer to fibers as the interface between matrix and fiber becomes strong, which also hinders crack propagation. The samples treated with cold water exhibited lower σ**UTS** and those treated with hot water exhibited the lowest strength due to the moisture absorbed at the interface which weakens the interface strength. Moisture absorption increases with temperature.

Prabhakar et al. [11] studied the effect of Nanomer I.28E on the mechanical properties of EGCN. In addition, the glass fiber was treated with silane and acid to check for the effects of both treatments. The results indicated that a combination of silane-treated glass fibers and Nanomer I.28E in the composite exhibited the highest σ**UTS** which was 130% compared to EGCN with untreated glass fiber and unmodified MMT particles. Prabhakar et al. [11] showed that any increase in an interfacial bond due to the addition of nanoclay led to increases in both σ**UTS** and hardness of the composite. The treatment of fibers and organic modifiers on MMT formed a strong interface.

2.2 Flexural properties

Haque et al. [26] reported 24% and 17% enhancement in flexural strength (σ**f**) and modulus at 1 wt% addition of Nanomer 1.28E in EGCN (shown **Figure 3a**). Kornmann et al. [17] reported 6% and 27% improvement in flexural modulus and σ**f** of EGCN at 10 wt% addition of ME-ODA (shown **Figure 3b**). The increase in σ**f** is linked to the existence of nano-silicate layers at the interface of the fiber, which might have improved interfacial properties. Another possible illustration is the fact that the compressive strength of epoxy is enhanced by the presence of the silicate layers so that it in turn enhances the bending strength of the laminate. Bozkurt et al. [18] reported 16% and 13% improvement in σ**f** and modulus at 6 wt% addition of OMMT (shown **Figure 3c**). It was observed from the fracture surface that fracture occurred along with the fiber-matrix interface and the fracture surface seems to have roughness indicating a strong interface. The laminate without clay showed a smooth fracture surface, which means the interface was weak. The increased flexural properties with the addition of various surface-modified nanoclays under different mixing conditions and making methods are given in **Table 1**.

Manfredi et al. [12] stated that the addition of Cloisite 10A in EGCN laminates caused the flexural modulus and strength to rise by 20% and 29%. The addition of Cloisite 30B did not cause any increment in the modulus of epoxy. It could be because of the collapse of clay particles, i.e. the particles were aggregated and the layers were not separated in the matrix. The modulus of clay nanoparticles is about 170 GPa (shown in **Figure 3d**). Therefore, when a strong bond is formed between matrix and clay it will result in an increased modulus of the laminate [51]. The increase in bending strength is attributed to the presence of silicate layers upon the glass fiber surface which improves the adhesion between the interface of matrix and glass fibers. The other possible reason for the improvement in the bending strength of laminate could be the increase in compressive strength of the epoxy. Shi and Kanny [19] reported that EGCN showed about 23% and 14% enhancement in modulus and strength at 3 wt% of Cloisite 30B. This enhancement is ascribed to the presence of intercalated silicate platelets of clay which interrupted the molecular motion of epoxy [52, 53]. The composites consisting of matrix processed by HSMT have shown 9.7% and 8.5% improvement in strength and modulus at 1 wt% (shown **Figure 3e**).

Sharma et al. [30] observed improvement in σf up to 5 wt% addition of nanoclay (shown **Figure 3f**). This increment is attributed to the presence of layered silicates on the glass fiber surface enhancing the adhesive bond between the epoxy matrix and glass fibers. In the range of 6 – 8 wt% of OMMT σf is reduced which was attributed to the agglomeration of OMMT in the EGCN. The uniform distribution and dispersion of silicate layers in epoxy resin are limited by the weight content of OMMT, when this content exceeds its percolation threshold (the ability of the liquid resin to pass through clay particles so that all the particles get wetted) there is a tendency to form particle aggregates [27]. The increased viscosity hinders the dispersion and favors the formation of agglomerates [54]. The fracture surface of GRE has shown that the fibers pulled out from the matrix had a smooth surface texture, whereas EGCN showed less fiber pullout with rough surfaces of fiber and matrix indicating the strong bond between fiber and epoxy and improved stress transfer between fiber and matrix. At 8 wt% clay addition, there were agglomerates formed fully in the EGCN [55].

At 40% and 60% volume of glass fiber reinforcement into epoxy-clay matrix, there was about 20% and 8% improvement in **σf** at 3 wt% of Nanomer I.30E (shown **Figure 3g**) [32]. This increment is attributed to the ability of the matrix to transfer the load to all the fibers. When nanoclay is not present in the matrix, it cannot transfer the load to all fibers, and thus crack propagates along with the matrix, and there will be low resistance to crack propagation. At low fiber volumes, i.e., at 40%, GRE exhibited interlaminar fracture as the crack propagated through the matrix between fiber layers and confined itself to layers near the top of the composite where the loading point is located so that the load was not transferred to all the layers, whereas EGCN exhibited translaminar fracture as the fiber layers break vertically at the loading point which requires more energy because the load is transferred to all the fiber layers [56].

With the increase in V_f of fiber to 60%, there was a reduction in the effect of nanoclay and both GRE and EGCN have failed predominantly in translaminar fracture mode which should occur only for EGCNs. This is because, at higher V_f of fibers, the fabric layers are well compacted to fit in the same volume of the composite, thereby the crimp zones present in the fabric will get interlocked with adjacent fabric layers, thus strengthening the interlaminar regions. Hence the crack propagation is resisted along interlaminar regions by the interlocked crimp zones and fracture occurs by rupture of glass fibers along the translaminar direction. These interlocks could resist interface shearing; thus, at higher V_f, crack propagation proceeds with the rupture of fiber fabric layers [32]. Srikanth I et al. [23] stated that At further higher fiber volumes, i.e., at >60%, fiber wetting became difficult, so there is a chance of failure by both mechanisms, i.e., interlaminar and translaminar crack propagation, thereby decreasing strength.

Jeyakumar et al. [33] stated that with the addition of Cloisite 93A into EGCN there was a significant improvement in flexural properties. With the addition of 1, 3, 5 wt% of Cloisite 93A, there was about 10.4%, 41.2%, and 52.3% increase in **σf** and also

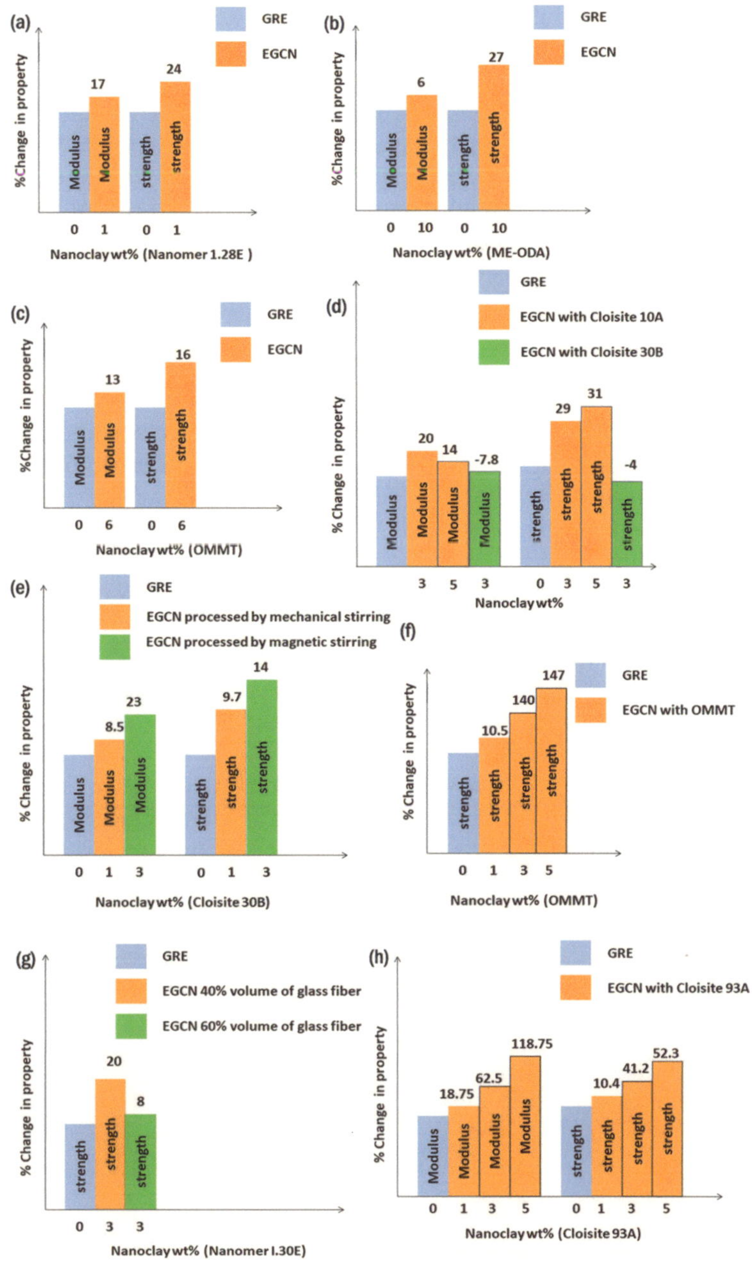

Figure 3.
Changes in flexural properties of EGCN'S at various conditions.

18.75%, 62.5%, and 118.75% improvement in flexural modulus. Beyond the 5 wt% addition of nanoclay, there was a decreasing trend (shown in **Figure 3h**). Najafi et al. [20] conducted experiments on EGCNs by adding pristine MMT and subjected some samples to hygrothermal conditions which consists of immersing the specimens in distilled water at 80°C for 10 weeks. The flexural curves for both neat GRE and EGCN exhibited linear behavior, EGCN subjected to hygrothermal conditions exhibited a gradual decrease in slope. At 3 and 5 wt% addition of MMT, there was about 8% and

12% improvement in flexural modulus, and 10.7% and 6.3% improvement in **σf** was observed. At 3 wt% of MMT addition, the properties were optimum. The samples treated by hygrothermal conditioning exhibited very poor flexural properties due to decreased interface bond strength caused by water absorption. Prabhakar et al. [11] stated that the addition of silane treated glass fiber in epoxy has resulted in improved flexural properties due to enhanced interface bonding between fiber and matrix compared to the composite reinforced with untreated fiber. The addition of Pristine MMT and Nanomer I.28E has not shown any considerable improvement but rather reduced the **σf**. There was about a 29% increase in **σf** of epoxy-silane treated fiber composite compared to epoxy-untreated fiber composite.

2.3 Fracture toughness

At 1 and 2 wt% addition of Nanomer I.28E, there was about 28% and 32% improvement in fracture toughness of clay-epoxy nanocomposite compared to NE, whereas EGCN exhibited about 20 and 23% improvement in fracture toughness for the same clay contents compared to GRE. Above 5 wt%, there was a decreasing tendency (shown **Figure 4a**) [26]. In the single edge notch bending test conducted by Bozkurt et al. [18], at 10 wt% addition of OMMT, the K_{IC} of EGCN improved by 5% but MMT did not show significant improvement (shown **Figure 4b**). The load applied is in the in-plane of the specimen. Therefore the fracture mechanism consisted of fiber-matrix debonding, fiber pullout, and fracture. The increased fracture toughness of the composites with the addition of various surface modified nanoclays under different mixing conditions and various making methods are given in **Table 1**.

Zulfli and Chow [27] stated that with the addition of nanoclay, K_{IC} improved. This improvement was ascribed to the strengthening of the interface between fiber and matrix by the presence of OMMT at the interface and increased resistance to crack propagation because of OMMT [55]. Swaminathan and Shivakumar [21] stated

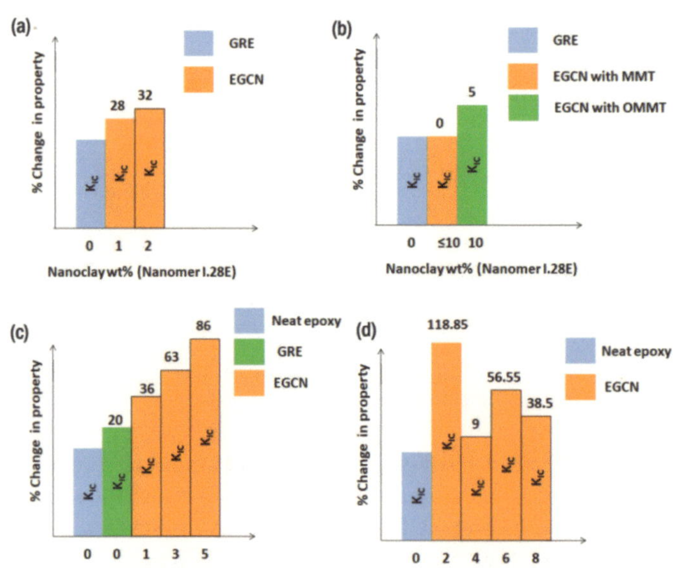

Figure 4.
Changes in fracture toughness of EGCN'S under various conditions.

that the major mechanism for increased toughness in composites was because of the deflection of the crack around clay tactoids. OMMT resists the crack from propagating because of which bowing and pinning of the crack take place [21]. The toughening effect of OMMT is limited by agglomeration. Tsai and Wu [22] reported a continuous decrease in Mode-I fracture toughness with the addition of nanoclay due to the brittleness induced in the composite which caused the crack to propagate at a faster rate, whereas pristine GRE composite exhibited ductile nature compared to EGCN with high clay content, so the crack propagation was slow and needed more energy for failure.

Jeyakumar et al. [33] reported that with the addition of Cloisite 93A into glass-epoxy, there was a conspicuous increase in fracture toughness of EGCN. For neat epoxy it was 0.9 MPa-$m^{1/2}$, for glass-epoxy it was 1.1 MPa-$m^{1/2}$. At 1, 3, and 5 wt% addition of nanoclay, the increase in fracture toughness of EGCN was about 36%, 63%, and 86% respectively compared to GRE. Beyond 5 wt% addition, there was a decreasing tendency (shown in **Figure 4c**). Therefore, it was concluded that the saturation limit is 5 wt% of nanoclay for the experimental conditions adopted by Jeyakumar et al. [33]. Senthil Kumar et al. [24] reported that with the addition of Cloisite 25A in EGCN, there was a considerable improvement in Mode-I fracture toughness of EGCN. At 2, 4, 6, and 8 wt% addition of nanoclay, there was about 118.85%, 9%, 56.55%, and 38.5% improvement in fracture toughness. Beyond 8 wt% addition, there was a decreasing trend (shown in **Figure 4d**). The increase in fracture toughness is attributed to the fiber bridging effect. At 10 wt% addition of nanoclay there was a decrease in the property, which is ascribed to the poor distribution of matrix between the fiber laminas.

2.4 Interlaminar shear strength (ILSS)

ILSS is a matrix dependent property, which means the strengthening of the matrix improves ILSS because the interface between the epoxy-clay matrix and the glass fiber becomes strong [57]. Therefore if the ILSS of the matrix is enhanced, then the ILSS of the composite also will get enhanced. The increase in ILSS of the composite is owing to the enhanced interfacial area between matrix and clay, the enhanced bond between resin and fiber, and the improved morphology of the matrix. The failure in ILSS mode is acknowledged as a critical mode of failure in FRP laminates. Thus there is a necessity to study the ILSS characteristics of the nanocomposites. It is proved that the shear strength of FRPs is remarkably enhanced with the incorporation of nanoclays [36]. EGCN with 1 and 2 wt% added Nanomer I.28E had shown 44% and 20% improvement in ILSS compared to GRE. The rough interface between the epoxy-fiber in fracture surface indicates a strong bond, whereas GRE and NE have shown a smooth interface which implies a weaker interface bond (shown **Figure 5a**) [26]. The increased ILSS of the composites with the addition of various surface modified nanoclays under different mixing conditions and making methods are given in **Table 1**. Bozkurt et al. [18] reported a decrease in ILSS of EGCN with the addition of MMT and OMMT. The ILSS of GRE is noted to be 32.7 MPa. But when the clay is added, it is observed that the laminate with the addition of clay reports a small decrease than when MMT is added; the decrease is high when OMMT is added. This decreasing trend is attributed to the creation of air voids in the interlaminar region while making the composite. The susceptibility to form voids in the interlaminar region is observed to be more when OMMT was added and further study is required to establish this phenomenon.

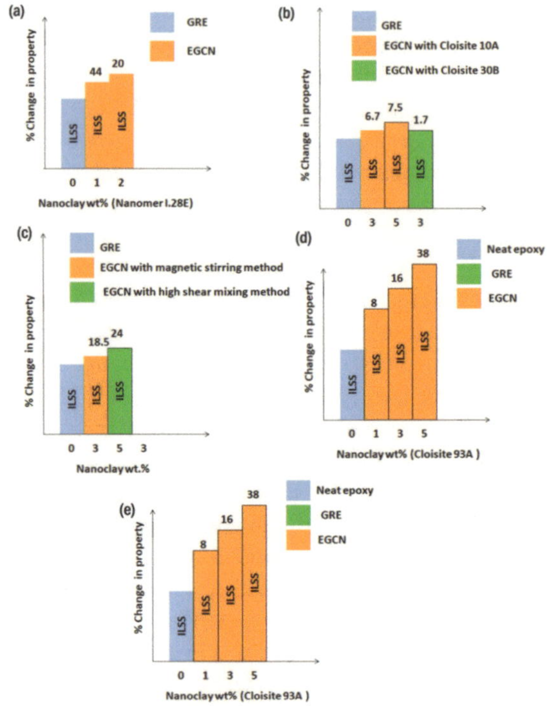

Figure 5.
Changes in ILSS of EGCN'S at various conditions.

The ILSS characteristics of GRE and EGCN with the addition of Cloisite 10A and Cloisite 30B were evaluated by Manfredi et al. [12]. There was a small increase of 7.5% in ILSS of EGCN with the addition of Cloisite 10A, but Cloisite 30B had no influence (shown **Figure 5b**). The trend of improvement with the addition of Cloisite 10A and decrease with the addition of Cloisite 30B was reported in the flexural properties section also. Laminates with Cloisite 10A have shown high flexural modulus and high **σf**. The morphologies of the composites indicated that the addition of Cloisite 30B had not provided strong adherence between matrix and fiber, but Cloisite 10A provided strong bonding between matrix and fiber. There is also a high attraction between Cloisite 10A and glass fiber surfaces since both are ceramic materials. The matrix without clay has shown a smooth and brittle surface at failure, whereas the matrix with nanoclay addition has shown a rough surface at failure which is also in line with the impact characteristics [12]. EGCN showed an 18.5% improvement in ILSS with the addition of 1 wt% of nanoclay by the magnetic stirring method. Above 1 wt%, there was a decreasing trend which is attributed to the aggregates of silicate tactoids and voids, whereas EGCN consists of a matrix processed by HSMT exhibited a 24% increase in ILSS, which might be attributed to the high shear force, which resulted in good dispersion of nanoclay platelets (shown **Figure 5c**) [19].

Jeyakumar et al. [33] reported that with the addition of Cloisite 93A, the ILSS of EGCN improved notably. At 1, 3, and 5 wt% addition of nanoclay in EGCN, there was about an 8%, 16%, and 38% increase in ILSS (shown **Figure 5d**). The presence of nanoclay brought about strong adhesion amongst nanoclay and epoxy matrix and in this manner enhanced the shear properties of the composites. Beyond 5 wt% the

ILSS started decreasing which might be due to the non-uniform scattering of nanoclay. Anni et al. [57] stated that with the addition of organically modified nanoclay into woven flax fiber reinforced epoxy, there was a rise in ILSS. Before reinforcing the fibers, some flax fibers were washed in distilled water, some treated with alkali solution, some with saline solution, and some others treated with nanoclay dispersed solution, to graft the nanoclay particles onto the flax fibers. The improvement in ILSS with the addition of these four kinds of treated fibers in ILSS was observed to be 8%, 10%, 17.9% compared to the composite reinforced only with water treated fibers.

Senthil Kumar et al. [24] reported that with the addition of Cloisite 25A into EGCN there was a significant increment in the ILSS property of EGCN. There was an increasing trend in the property up to 2 wt% addition of nanoclay, after that, it started decreasing. At 2 wt% of nanoclay addition, there was about a 70% increase in ILSS of EGCN (shown **Figure 5e**). ILSS mainly depends on matrix behavior if the matrix is tough, the ILSS is increased. The addition of nanoclay makes the matrix tough because the crack propagation is hindered by the clay platelets and the stress distributed to the fibers will be uniform as the interface becomes stronger. At 10 wt% addition of nanoclay, the ILSS decreased by 3% compared to GRE. Lim et al. [58] showed that the geometry of the interface between epoxy-nanoclay platelets may also influence ILSS.

2.5 Impact strength

The impact strength of the composite depends mainly on the strength of the matrix and the ability of the fiber matrix to withstand the impact loads. At 5 wt% addition of Cloisite 10A, the EGCN has exhibited a 23% improvement in impact strength; this improvement is attributed to the creation of a complex path for the fracture propagation, as the layered silicate platelets hinder the extension of microcracks created in the matrix (shown **Figure 6a**) [59]. The increase in the strength of the fiber-matrix interface has decreased the resistance to impact force. Manfredi et al. [12] stated that the failure strength of EGCN depends on two factors, one being the tortuous path formed by clay platelets, and another being the fiber-matrix interface strength. The well-dispersed nanoclay platelets hinder crack propagation by diverting the crack to a longer path or splitting it into sub cracks that require more energy, whereas a strong fiber-matrix interface reduces the impact resistance. The laminates were made with low fiber content hence the properties of the laminate are mainly dependent on the matrix behavior. An improvement in the impact characteristics of the nanocomposite with no glass fiber reinforcement was observed. The enhancement in the impact characteristics was observed for laminates with nanocomposite matrix irrespective of the clay type [12].

Shi and Kanny [19] carried out an Izod impact test at a high strain rate to study the impact characteristics of EGCN. When the matrix incorporated into the laminate was processed by magnetic stirring, the impact strength of the laminate was noticed to be decreasing with the addition of Cloisite 30B. A sudden decrease in impact strength of 27% is observed for the laminate at 1 wt% clay; further addition of clay did not affect impact strength (shown **Figure 6b**). The sudden decrease at 1 wt% clay is attributed to the agglomeration and air voids in the matrix, Siddiqui et al. [60] addressed the same finding, whereas 44.9% improvement in the impact strength.

At 1 wt% nanoclay was observed when the laminate prepared was incorporated with a matrix processed by HSMT [19]. The changes in the impact strength of the composites with the addition of various surface modified nanoclays under different mixing conditions and making methods are given in **Table 1**.

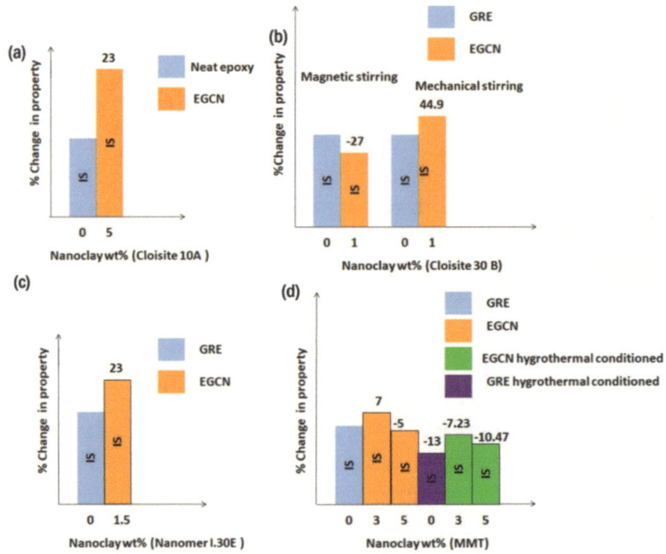

Figure 6.
Changes in impact strength of EGCN'S under various conditions.

Zulfli and Chow [27] reported that the impact strength of the laminates with Nanomer 1.28E incorporated in the matrix exhibited a higher value compared to GRE. This improvement in the impact characteristics was ascribed to strong adhesion between Nanomer 1.28E and epoxy which implies that the resin has wetted all layers of the nanoclay particles. This, therefore, enhances the energy required to debond the fiber and matrix due to the strong bond. Yasmin et al. [49] stated that the enhanced impact strength of the laminate is because of the complex path for cracks to propagate through the matrix. The OMMT and glass fiber provides a synergistic increment to the impact characteristics. OMMT at the fiber matrix interface acts as an interfacial modifier while the stress transfer from the matrix to fiber gets enhanced through clay particles; thus as the clay content at the fiber matrix interface increases, higher stress levels can be taken by the composite because of which better characteristics were attained [40]. But the content of clay that can be added to the epoxy is limited by the agglomeration and air voids that are formed while mixing the clay into the resin. Rafiq Ahmad et al. [41] added Nanomer I.30E into EGCN to evaluate its effect on the impact strength of EGCN. The laminates were stroked with low-speed impact forces ranging between 10 and 50 J. The optimum property was obtained at 1.5 wt% of nanoclay addition with 23% improvement in the maximum load required to damage the specimen and 11% improvement in stiffness. Also, a notable decrease in physical damage was observed for EGCN compared to GRE (shown **Figure 6c**).

Najafi et al. [20] studied the effect of the addition of pristine MMT into EGCN on impact strength. To study the effect of hygrothermal aging, some EGCN specimens were immersed in distilled water at 80°C for 10 weeks. At 3 wt% nanoclay addition there was about a 7% increase in impact strength for EGCN. At 5 wt%, the impact strength reduced nearly by 5% compared to the value obtained at 3 wt%, and this decrease was attributed to agglomerates. Also, the brittleness of EGCN increased with the addition of nanoclay, causing the energy absorption to decrease [59]. The 3 wt% and 5 wt% nanoclay added EGCN subjected to hygrothermal conditioning exhibited a 7.23% and 10.47% decrease in impact strength compared to the control specimen which was dry GRE. The conditioned GRE exhibited about 13% decrease compared to dry GRE, whereas for 3 and 5 wt% added, conditioned EGCN exhibited about 14% and 8% increase compared to conditioned GRE (shown **Figure 6d**). In both dry and conditioned states, the 3 wt% added EGCN's exhibited good impact strength compared to the control specimen. Prabhakar et al. [11] stated that EGCN reinforced with acid treated glass fiber and MMT exhibited the highest impact strength out of all the composites made using neat glass fiber, silane treated glass fiber, and acid treated glass fiber, MMT, and Nanomer I.28E. Neat GRE exhibited the second highest impact strength value. The next highest impact strength was exhibited by EGCN with silane treated fiber and Nanomer I.28E. Compared to neat GRE the former one was 2% superior in property and the latter one is 2% inferior in the property. Prabhakar et al. [11] stated that a decrease in impact strength was compensated by an improvement in hardness of composites added with Nanomer I.28E and silane treated fiber, because the increase in hardness increases the brittleness, thereby reducing the energy absorption capability.

3. Conclusion

After reviewing the existing literature available on EGCNs reinforced with surface modified nanoclays, it is clear that the interfacial bond between reinforced fibers and the matrix is enhanced which resulted in enhancement in the mechanical properties of the composite. The enhanced fiber-matrix interface strength is due to good adhesion between clay platelets and epoxy allowing better stress transfer to all the fibers.

Conflict of interest

The authors have declared no conflict of interest.

Author details

Shanti Kiran Zhade[1], Syam Kumar Chokka[2]*, V. Suresh Babu[3] and K.V. Sai Srinadh[3]

1 RGUKT, Basar, India

2 Ellenki College of Engineering and Technology, India

3 NIT, Warangal, India

*Address all correspondence to: chokka.syamkumar@gmail.com

References

[1] Kaczmar JW, Pietrzak K, Wlonsinski W. The production and application of metal matrix composite materials. Journal Material Processing Technology. 2000;**106**:58-67. DOI: 10.1016/S0924-0136(00)00639-7

[2] Miracle DB. Metal matrix composites– From science to technological significance. Composites Science and Technology. 2005;**65**:2526-2540. DOI: 10.1016/j.compscitech.2005.05.027

[3] Mortensen A, Llorca J. Metal matrix composites. Annual Review of Materials Research. 2010;**40**:243-270. DOI: 10.1146/annurev-matsci-070909-104511

[4] Mc Garry FJ. Polymer composites. Annual Review of Materials Science. 1994;**24**:63-82. DOI: 10.1146/annurev.ms.24.080194.000431

[5] Saheb DN, Jog JP. Natural fiber polymer composites: A review. Advances in Polymer Technology. 1999;**18**: 351-363. DOI: 10.1002/(SICI)1098-2329(199924)18:4<351:AID-ADV6>3.0.CO;2-X

[6] Matabola KP, Vries AR De, Moolman FS, Luyt AS. Single polymer composites: A review. Journal of Materials Science 2009;44:6213-6222. DOI: 10.1007/s10853-009-3792-1

[7] Ku H, Wang H, Pattarachaiyakoop N, Trada M. A review on the tensile properties of natural fiber reinforced polymer composites. Composites Part B: Engineering. 2011;**42**:856-873. DOI: 10.1016/j.compositesb.2011.01.010

[8] Dhingra AK. Metals Replacement by Composites. The Journal of The Minerals, Metals & Materials Society. 1986;**38**:17. DOI: 10.1007/BF03257887

[9] Chakraverty AP. Some aspects of evaluation of GFRP composite [thesis]. Rourkela: National Institute of Technology; 2015

[10] Roco MC. Nanoparticles and Nanotechnology Research. Journal of Nanoparticle Research. 1999;**1**:1-6. DOI: 10.1023/A:1010093308079

[11] Prabhakar K, Debnath S, Anwar M, Palanikumar K. Experimental analysis on the effect of surface treatment of glass fibers & nanoclay on mechanical properties of glass fiber reinforced polymer nanocomposites. Conf. Series: Materials Science and Engineering. 2019;**495**:012091. DOI: 10.1088/1757-899X/495/1/012091

[12] Manfredi LB, De Santis H, Vázquez A. Influence of the addition of montmorillonite to the matrix of unidirectional glass fibre/epoxy composites on their mechanical and water absorption properties. Composites Part A: Applied Science and Manufacturing. 2008;**39**:1726-1731. DOI: 10.1016/j.compositesa.2008.07.016

[13] Lan T, Kaviratna PD, Pinnavaia TJ. Mechanism of clay tactoid exfoliation in epoxy-clay nanocomposites. Chemistry of Materials. 1995;7:2144-2215. DOI: 10.1021/cm00059a023

[14] Wang Z, Pinnavaia TJ. Hybrid organic-inorganic nanocomposites: Exfoliation of magadiite nanolayers in an elastomeric epoxy polymer. Chemistry of Materials. 1998;**10**:1820-1826. DOI: 10.1021/cm970784o

[15] Lagaly G. Interaction of alkylamines with different types of layered compounds. Solid State Ionics. 1986;**22**:43-51

[16] Alexandre M, Dubois P. Polymer-layered silicate nanocomposites: preparation, properties and uses of a new class of materials. Materials Science and Engineering. 2000;**28**:1-63. DOI: 10.1016/S0927-796X(00)00012-7

[17] Kornmann X, Rees M, Thomann Y, Necola A, Barbezat M, Thomann R. Epoxy-layered silicate nanocomposites as matrix in glass fibre-reinforced composites. Composites Science and Technology. 2005;**65**:2259-2268. DOI: 10.1016/j.compscitech.2005.02.006

[18] Bozkurt E, Kaya E, Tanolu M. Mechanical and thermal behavior of non-crimp glass fiber reinforced layered clay/epoxy nanocomposites. Composites Science and Technology. 2007;**67**:3394-3403. DOI: 10.1016/j.compscitech.2007.03.021

[19] Shi Y, Kanny K, Jawahar P. Hybrid nanocomposites: Processing and properties. Advanced Composite Materials. 2009;**18**:365-379. DOI: 10.1163/156855109X434757

[20] Najafi M, Darvizeh A, Ansari R. Effect of nanoclay addition on the hygrothermal durability of glass/epoxy and fiber metal laminates. Fibers and Polymers. 2018;**19**:1956-1969. DOI: 10.1007/s12221-018-8235-7

[21] Swaminathan G, Shivakumar K. Thermomechanical and fracture properties of exfoliated nanoclay nanocomposites. Journal of Reinforced Plastics and Composites. 2011;**30**:256-268. DOI: 10.1177/0731684410393052

[22] Tsai J-L, Wu M-D. Organoclay effect on mechanical responses of glass/epoxy nanocomposites. Journal of Composite Materials. 2008;**42**:553-568. DOI: 10.1177/0021998307087514

[23] Srikanth I, Kumar S, Singh V, Rangababu B. Effect of carbon nanofibre addition on the mechanical properties of different V_f carbon-epoxy composites. Bulletin of Materials Science. 2015;**38**: 309-317

[24] Senthil Kumar MS, Chithirai Pon Selvanb M, Sampathc PS, Raja K, Balasundaram K. Influence of nanoclay on interlaminar shear strength and fracture toughness of glass fiber reinforced nanocomposites Influence of nanoclay on interlaminar shear strength and fracture toughness of glass fiber reinforced nanocomposites. IOP Conference Series: Materials Science and Engineering. 2018;**346**:012081. DOI: 10.1088/1757-899X/346/1/012081

[25] Steinmann W, Saelhoff AK. Essential properties of fibres for composite applications. Rana S, Fangueiro R, editors. Fibrous and Textile Materials for Composite Applications. Textile Science and Clothing Technology. Singapore. Springer; 2016. p. 39-73. DOI: 10.1007/978-981-10-0234-2_2

[26] Haque A, Shamsuzzoha M, Hussain F, Dean D. S2-glass/epoxy polymer nanocomposites: Manufacturing, structures, thermal and mechanical properties. Journal of Composite Materials. 2003;**37**:1821-1838. DOI: 10.1177/002199803035186

[27] Zulfli M, Chow W. Mechanical and thermal behaviours of glass fiber reinforced epoxy hybrid composites containing organo-montmorillonite. Malaysian Polymer Journal. 2012;7:1-8

[28] Kanny K, Mohan TP. Resin infusion analysis of nanoclay filled glass fiber laminates. Composites Part B: Engineering. 2014;**58**:328-334. DOI: 10.1016/j.compositesb.2013.10.025

[29] Gurusideswar S, Velmurugan R. Strain rate sensitivity of glass/epoxy composites with nanofillers. Materials

and Design. 2014;**60**:468-478. DOI: 10.1016/j.matdes.2014.03.065

[30] Sharma B, Mahajan S, Chhibber R, Mehta R. Glass fiber reinforced polymer-clay nanocomposites: processing, structure and hygrothermal effects on mechanical properties. Procedia Chemistry. 2012;**4**:39-46. DOI: 10.1016/j.proche.2012.06.006

[31] Gurusideswar S, Velmurugan R. High Strain Rate Sensitivity of Glass/Epoxy/Clay Nanocomposites. Losboa. Portugal: Proceedings of 10th International Conference on Composite Science and Technology (ICCST '10); 2-4 September 2015. pp. 1-12

[32] Krushnamurty K, Rasmitha D, Srikanth I, Ramji K, Subrahmanyam C. Effect of Volume Fraction of Fibre on the Mechanical Properties of Nanoclay Reinforced E-Glass-Epoxy Composites. International Journal of Materials and Metallurgical Engineering. 2016;**10**:98-101. DOI: scholar.waset.org/1999.2/10003722

[33] Jeyakumar R, Sampath PS, Ramamoorthi R, Ramakrishnan T. Structural, morphological and mechanical behaviour of glass fibre reinforced epoxy nanoclay composites. International Journal of Advanced Manufacturing Technology. 2017;**93**:527-535. DOI: 10.1007/s00170-017-0565-x

[34] Kini UA, Shettar M, Sathyashankara Sharma PH, GM C, Hegde A, Siddhartha D. Effect of hygrothermal aging on the mechanical properties of nanoclay- glass fiber-epoxy composite and optimization using full factorial design. Materials Research Express. 2019;**6**:1-13. DOI: 10.1088/2053-1591/ab0d68

[35] Jawahar P, Balasubramanian M. Influence of nanosize clay platelets on the mechanical properties of glass fiber reinforced polyester composites. Journal of Nanoscience and Nanotechnology. 2006;**6**:3973-3976. DOI: 10.1166/jnn.2006.656

[36] Wang X, Zhao X, Wu Z, Zhu Z, Wang Z. Interlaminar shear behavior of basalt FRP and hybrid FRP laminates. Journal of Composite Materials. 2016;**50**:1073-1084. DOI: 10.1177/0021998315587132

[37] Lin JC, Chang LC, Nien MH, Ho HL. Mechanical behavior of various nanoparticle filled composites at low-velocity impact. Composite Structures. 2006;**74**:30-36. DOI: 10.1016/j.compstruct.2005.03.006

[38] Basara C, Yilmazer U, Bayram G. Synthesis and characterization of epoxy based nanocomposites. Journal of Applied Polymer Science. 2005;**98**:1081-1086. DOI: 10.1002/app.22242

[39] Taylor AC, Kinloch AJ. Mechanical and fracture properties of epoxy/inorganic micro-and nano-composites. Journal of Materials Science Letters. 2003;**22**:1439-1442. DOI: 10.1023/A

[40] Mishra S, Mohanty AK, Drzal LT, Misra M, Parija S, Nayak SK, et al. Studies on mechanical performance of biofibre/glass reinforced polyester hybrid composites. Composites Science and Technology. 2003;**63**:1377-1385. DOI: 10.1016/S0266-3538(03)00084-8

[41] Rafiq A, Merah N, Boukhili R, Al-qadhi M. Impact resistance of hybrid glass fiber reinforced epoxy/nanoclay composite. Polymer Testing. 2017;**57**: 1-11. DOI: 10.1016/j.polymertesting.2016.11.005

[42] Okoli O, Smith G. High strain rate characterization of a glass/epoxy composite. Journal of Composites

Technology and Research. 2000;**22**:3. DOI: 10.1520/CTR10619J

[43] Fereshteh-Saniee F, Majzoobi GH, Bahrami M. An experimental study on the behavior of glass-epoxy composite at low strain rates. Journal of Materials Processing Technology. 2005;**162-163**:39-45. DOI: 10.1016/j.jmatprotec.2005.02.011

[44] Shokrieh MM, Omidi MJ. Tension behavior of unidirectional glass/epoxy composites under different strain rates. Composite Structures. 2009;**88**:595-601. DOI: 10.1016/j.compstruct.2008.06.012

[45] Hussain F. Polymer-matrix nanocomposites, processing, manufacturing, and application: An overview. Journal of Composite Materials. 2006;**40**:1511-1575. DOI: 10.1177/0021998306067321

[46] Li X, Zhan Z, Peng G, Wang W. A new method for preparing completely exfoliated epoxy/clay nanocomposites: Nano-disassembling method. Polymer Bulletin. 2011;**67**:719-727. DOI 10.1007/s00289-011-0496-x

[47] Withers GJ, Yu Y, Khabashesku VN, Cercone L, Hadjiev VG, Souza JM, et al. Improved mechanical properties of an epoxy glass-fiber composite reinforced with surface organomodified nanoclays. Composites Part B: Engineering. 2015;**72**:175-182. DOI: 10.1016/j.compositesb.2014.12.008

[48] Brown KA, Brooks R, Warrior NA. The static and high strain rate behaviour of a commingled E-glass/polypropylene woven fabric composite. Composites Science and Technology. 2010;**70**:272-283. DOI: 10.1016/j.compscitech.2009.10.018

[49] Yasmin A, Luo JJ, Abot JL, Daniel IM. Mechanical and thermal behavior of clay/epoxy nanocomposites. Composites Science and Technology. 2006;**66**:2415-2422. DOI: 10.1016/j.compscitech.2006.03.011

[50] Yasmin A, Abot JL, Daniel IM. Processing of clay/epoxy nanocomposites by shear mixing. Scripta Materialia. 2003;**49**:81-86. DOI: 10.1016/S1359-6462(03)00173-8

[51] Sinha Ray S, Okamoto M. Polymer/layered silicate nanocomposites: A review from preparation to processing. Progress in Polymer Science. 2003;**28**:1539-1641. DOI: 10.1016/j.progpolymsci.2003.08.002

[52] Wu CL, Zhang MQ, Rong MZ, Friedrich K. Tensile performance improvement of low nanoparticles filled-polypropylene composites. Composites Science and Technology. 2002;**62**:1327-1340. DOI: 10.1016/S0266-3538(02)00079-9

[53] Mittal V. Polymer layered silicate nanocomposites: A review. Materials. 2009;**2**:992-1057. DOI: 10.3390/ma2030992

[54] Karippal JJ, Murthy HNN, Rai KS, Sreejith M, Krishna M. Study of mechanical properties of epoxy/glass/nanoclay hybrid composites. Journal of Composite Materials. 2011;**45**:1893-1899. DOI: 10.1177/0021998310389087

[55] Khan SU, Munir A, Hussain R, Kim JK. Fatigue damage behaviors of carbon fiber-reinforced epoxy composites containing nanoclay. Composites Science and Technology. 2010;**70**:2077-2085. DOI: 10.1016/j.compscitech.2010.08.004

[56] Wichmann MHG, Sumfleth J, Gojny FH. Glass-fibre-reinforced composites with enhanced mechanical and electrical properties–Benefits and limitations of a nanoparticle modified

matrix. Vol. 732006. pp. 2346-2359. DOI: 10.1016/j.engfracmech.2006.05.015

[57] Wang A, Xia D, Xian G, Li H. Effect of nanoclay grafting onto flax fibers on the interfacial shear strength and mechanical properties of flax/epoxy composites. Polymer Composites. 2019;**40**:1-11. DOI: 10.1002/pc.25210

[58] Lim JW, Hassan A, Rahmat AR, Wahit MU. Morphology, thermal and mechanical behavior of polypropylene nanocomposites toughened with poly (Ethylene-Co-Octene). Polymer International. 2006;**55**:204-215. DOI: 10.1002/pi.1942

[59] Khan SU, Iqbal K, Munir A, Kim JK. Quasi-static and impact fracture behaviors of CFRPs with nanoclay-filled epoxy matrix. Composites Part A: Applied Science and Manufacturing. 2011;**42**:253-264. DOI: 10.1016/j.compositesa.2010.11.011

[60] Siddiqui NA, Woo RSC, Kim J, CCK L. Mode I interlaminar fracture behavior and mechanical properties of CFRPs with nanoclay-filled epoxy matrix. Composites Part A: Applied Science and Manufacturing. 2007;**38**:449-460. DOI: 10.1016/j.compositesa.2006.03.001

Chapter 3

Fiber Inclusions-Based Epoxy Composites and Their Applications

Nassima Radouane and Abdelkrim Maaroufi

Abstract

Because of their low cost, lightweight, easy production methods, and design flexibility, polymer-based composites are widely employed in a wide range of applications. Because of its high specific strength, superior mechanical characteristics, super adhesiveness, heat and solvent resistance, and so on, epoxy polymer or polyepoxide represent a significant majority of matrix composites. As a result, fiber fillers-reinforced epoxy resin composites have been investigated for a variety of applications, including high-tech in the ballistic, aircraft, automobile, construction, and sports sectors. In this chapter, the manufacturing procedures of fiber-reinforced epoxy composites have been described. Different categories of fiber are used as fillers in an epoxy matrix and their morphology is discussed as a function of the obtained properties.

Keywords: carbon fiber, glass fiber, plant fiber, epoxy composites, applications

1. Introduction

In recent decades, our societies have been confronted with climatic disturbances and resource use, leading to the degradation of ecosystems. In order to combat these threatening changes, the international community is committed to finding new ways of producing and creating value, including light-weighting structures and the valorization of lignocellulosic biomass as possible solutions towards sustainable innovation [1–3]. Indeed, light-weighting implies a reduction in production energy, raw materials produced and materials to be managed at the end of life [4, 5]. The reduction in mass also leads to a reduction in the energy consumption of means of transport and their emissions of pollutants. For these reasons, sandwich structures are increasingly used instead of monolithic structures in various applications, thanks to their lightness, their mechanical performance in bending and their thermal, vibratory, and acoustic features.

Because of their good features, they may be found in vital industries such as aeronautics, automotive, sports, marine, and construction. These properties include high mechanical strength and stiffness, high-impact resistance, low weight, corrosion resistance, and low maintenance costs [6]. Traditionally, composite materials are reinforced with synthetic fibers such as glass, carbon, aramid, or ceramic fibers. These fibers are used because of their strength, stiffness, low moisture absorption,

 IntechOpen

and good compatibility with polymer resins. Glass fibers are the most commonly used because of their low cost, ease of production, and specific mechanical characteristics.

The epoxy matrix combined with rigid fiber allows for the creation of building materials with high stiffness and strength. Given the variety of technical and material options, developing a composite material necessitates taking into account the chemical and physical interactions between all components [7]. As a result, the effects of production processes, fiber reinforcement type, and reactive or nonreactive modifiers on the characteristics of epoxy composites remain intriguing study issues. Many different types of synthetic and natural fibers are used to strengthen the epoxy matrix, including glass, carbon, basalt, aramid, ramie, hemp, jute, and flax [8].

In this chapter, a detailed description of epoxy polymer was represented. Moreover, various fiber types such as glass, carbon, and plant materials. In addition, some fabrication procedures of epoxy reinforced fiber composite are reported. Furthermore, a representation of some applications was described as well as the coming challenges.

2. Thermosetting organic matrix: Epoxy resin

Epoxy resin is a thermosetting polymer. It comprises two parts: an epoxy base catalyst and an amine-containing hardener (-NH_2 or -NH). During cross-linking, each hydrogen atom in the amine group opens the epoxy ring and produces a polymer chain (**Figure 1**) [9]. The glass transition temperature denoted T_g, increases with the rate of crosslinking. Thanks to its 3D polymeric structure and high phase change temperature, epoxy phase change temperature, epoxy achieves good mechanical and thermal properties [10].

Table 1 shows the advantages and disadvantages of epoxy thermoset resin. Compared to thermoplastic resins, epoxy resins are more brittle on impact due to their susceptibility to cracking. According to Vieille et al. [12], the impact response of thermoset matrix composites has some weak points:

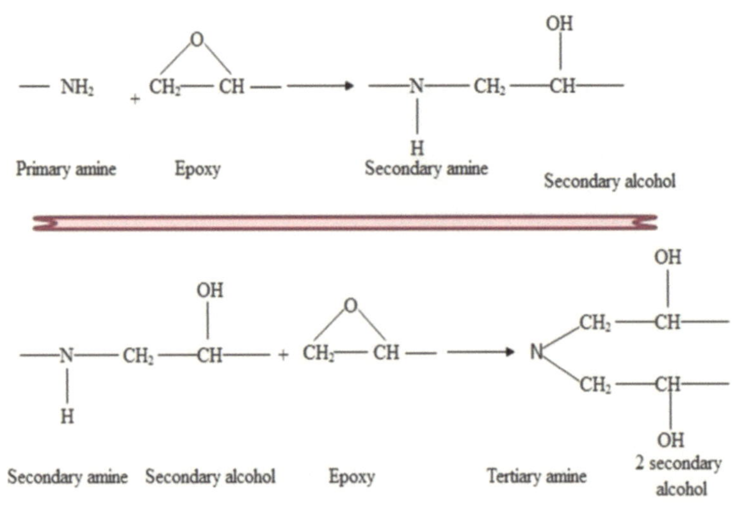

Figure 1.
Main chemical reactions taking place during the curing of an epoxy resin.

Thermosetting epoxy resin	
Benefits	**Disadvantages**
Mechanical properties superior to those of polymers of the same family (tensile, compression, aging, etc.);	Preservation at low temperature (for prepregs or single-component resin);
Good temperature resistance from 150–190°C approximately;	Long curing time;
Excellent chemical resistance;	High cost (about 5 times more than polyester resin);
Low molding shrinkage (0.5 to 1%);	Need to take precautions during implementation;
Good wettability of the reinforcements;	Sensitivity to cracking.
Bonne adhérence aux matériaux métalliques.	—

Table 1.
Advantages and disadvantages of epoxy resin [11].

- For the same impact energy, the delamination area due to impact is higher compared to thermoplastic matrix composites;
- For impact energy of 25 J, the epoxy reinforced structure is perforated while the maximum displacement of the thermoplastic composite structure is around 11.4 mm;
- For the same level of impact energy, the ratio of dissipated energy to impact energy of the epoxy matrix composite is higher than the ratio obtained for the thermoplastic matrix composite. The risk of perforation of the structure increases with the increase of this ratio.

Moreover, a high brittleness of this family of resins is also the cause of the pseudoplastic behavior of the composite. Upon impact, the opening of intralaminar and interlaminar cracks is triggered. At the same time, epoxy debris forms and blocks the closure of the cracks after impact, which is impact, which is unfavorable to the impact resistance of the composite [13].

3. Fiber materials and types

Composite materials are categorized based on their content, which includes the base material (matrix) and the filler material. A matrix or binder material is the basic material that binds or retains the filler material in structures, whereas filler material is present in the form of sheets, pieces, particles, fibers, or whiskers of natural or synthetic material. Composite based fibers are categorized into three major groups based on their structure, as shown in **Figure 2**.

3.1 Epoxy resins reinforced with glass fibers

Fiber or particulate inclusions in epoxy matrix with different types and shapes are studied by many researchers to characterize their mechanical, electrical, thermal

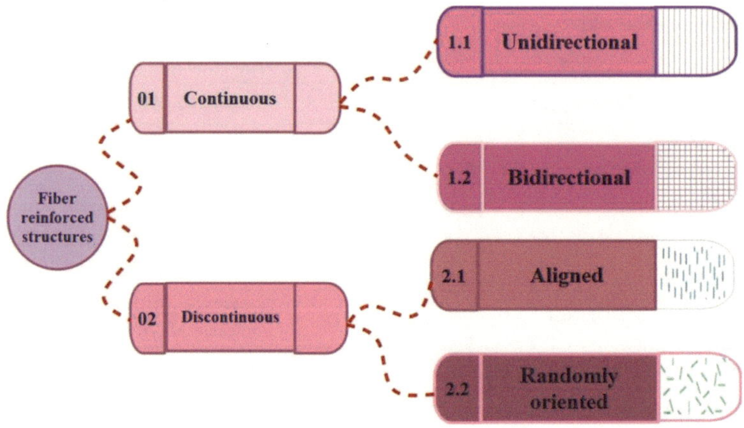

Figure 2.
Composites structure types.

and so on properties [14–17]. Glass fibers are the most often used synthetic fibers because of their high strength and durability, thermal stability, impact resistance, chemical, friction, and wear qualities. However, machining glass fiber-reinforced polymers (GFRPs) using traditional machining techniques is generally slow, difficult, and results in lower tool life [18]. They are easily made from raw material, which is readily available in an almost limitless supply. There are numerous types of GFs that are often utilized in GFRP composites, depending on the raw materials used and their quantities in fabrication (see **Figure 3**). GFs also have the disadvantage of being disposed of at the end of their useful life [19]. Glass fiber reinforced polymer

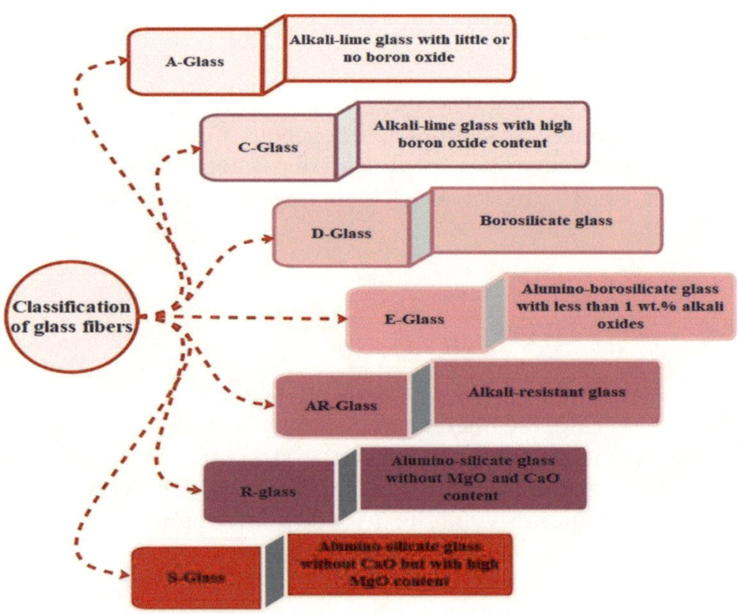

Figure 3.
Glass classification.

Fiber glass type	Manufacturing procedure	Composite structure type	Reference
E-Glass fiber	Hand lay-up technique	—	[21]
E-Glass fiber	Hand lay-up	Randomly oriented	[22]
Glass Fabrics	hand layup vacuum bagging	—	[23]
C-Glass fiber	Hand lay-up	Randomly oriented	[24]
E-Glass fiber	Open layup molding	—	[25]
D-Glass fiber	Hand lay-up techniques	Randomly oriented	[26]

Table 2.
Example of composites-based epoxy reinforced glass fibers.

composites were created using various production technologies and are widely employed in a variety of applications [20]. Because of their superior mechanical qualities, glass fiber reinforced composites have received more attention in recent years. Glass fibers have excellent features such as high strength, flexibility, stiffness, durability, and so on. The characteristics of GFRP composites improved when the amount of glass fiber was increased. The mechanical and thermal properties of different polymer composites reinforced with glass fiber when exposed to mechanical stress are been listed in the following **Table 2**.

3.2 Epoxy resins reinforced with carbon fibers

Carbon fibers were first used in 1880 by T. Edison as a filament in lamps. From 1960 onwards, research was directed towards the development of high modulus and high strength carbon fibers. The carbon fibers are more required in applications that need more stiffness. Carbon fiber-reinforced polymer (CFRP) composites have extensive uses in aircraft, automotive, sports, and a variety of other sectors. In the literature, many other fillers type such as particulate and fiber fillers [27, 28]. In general, carbon fibers can be categorized by their mechanical properties, manufacturing methods, application field, precursor, fiber materials, final heat treatment temperature, and their function.

3.3 Epoxy resins reinforced with plant fibers

Nowadays, industrial businesses are concentrating on providing environmentally friendly products, and the globe is moving towards sustainable development. Because of their biodegradability, natural fibers are employed in the production of such eco-friendly products. The key causes influencing the rising use of natural fiber-reinforced composites are increased awareness of concerns such as pollution, waste of raw materials and energy, and depletion of petroleum reserves (FRC). Due to their lightness, mechanical performance, ability to integrate functions, physical–chemical resistance, and ease of processing, composite materials have made considerable progress in terms of volume and have dominated practically all sectors. Such as wood fiber which are transformed using the steam explosion process and are treated at various steam pressures. Because of the increased explosion pressure, the fiber's affinity for water, mechanical characteristics, and dissolving ability in caustic solution diminish after the steam explosion [29].

4. Fabrication methods

The preferred procedure is determined mostly by the resin used, the length of the fibers, the required qualities of the composite material, and the production run and rate (**Figure 4**) [30].

- Contact molding: It is a procedure for short series. On a waxed mold, layers of catalyzed resin and layers of cloth are alternatively deposited. After applying the resin to the reinforcement with a brush, it is debulked with a roller. This process is continued until the appropriate number of layers has been attained. This method is easy and economical, and it allows for the fabrication of pieces of any form and size with a nice surface look on the mold side. The component produced by this approach is heavily reliant on the molder's competence.

- Vacuum molding: This is a method for producing medium-sized series. Following the placement of the reinforcements and resin on a waxed hard mold, a waterproof membrane is applied to the whole structure. A vacuum established between the mold and the membrane by a vacuum pump allows the resin to be distributed and debulked. The suction is kept up until the resin is completely dry.

- Low-pressure liquid resin injection molding: This method is also known as RTM (Resin Transfer Molding). It entails inserting the reinforcements into a hard, waxed, two-sided mold. The catalyzed resin is then pumped into the mold at pressures ranging from 0.1 to 0.4 MPa.

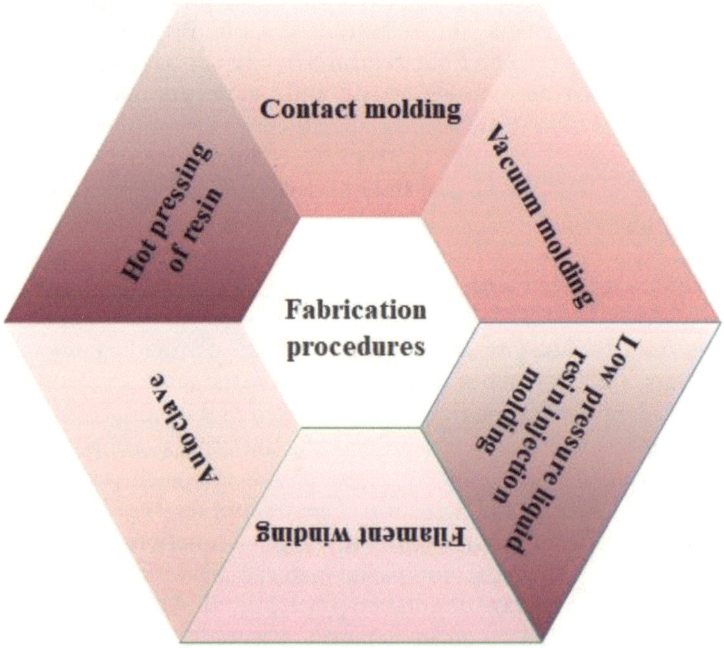

Figure 4.
Different procedures of fabrication for polymer composites-based fibers.

- Filament winding: This automated method is intended for the high-pressure molding of high-performance innovative parts like tubes and fluid storage tanks. The method entails dipping rovings in a bath of catalyzed resin before wrapping them around a mandrel. There are three forms of winding based on the speed of movement of the roving in proportion to the angle and speed of rotation of the mandrel: circumferential winding, polar winding, and helical winding.

- Autoclave: Bag molding is another name for autoclave molding. It entails compacting the reinforcement and resin on a stiff mold, then passing it through an elastic, flexible membrane to form a tight bag with the tooling. The mold is put in a confined chamber with a few megapascals of internal pressure. This pressure is given to the membrane by a fluid (air, water, nitrogen, or steam), which aids in resin polymerization.

- Simultaneous spray molding: This method evolved from the contact molding procedure. Molding is done by spraying chopped fibers combined with catalyzed resin onto a waxed mold at the same time. A roller is then used to condense the sprayed layer and eliminate bubbles. This technology allows for the low-cost production of medium and large parts as well as basic forms. However, because this form of molding solely employs chopped fibers, the pieces produced have poor mechanical qualities.

According to studies, the constraints of each technique and the production parameters employed during composite processing might induce undesired internal flaws into the material, such as bubbles or cavities, poor or rich areas, delamination, shrinkage, and so on [30]. As a result, these flaws can compromise the mechanical characteristics of composite materials. Several investigations on carbon and glass reinforced composites have been conducted. Liu et al. investigated the influence of autoclave pressure cycling on the porosity of a [0/90]3 s carbon/epoxy cross-linked composite. In comparison to tensile strength and modulus, they demonstrated a considerable sensitivity to porosity in the interlaminar shear strength and flexural parameters of the composites. They discovered that when the porosity inside the composites is less than 4%, the interlaminar shear strength reduces by roughly 8% for every 1% increase in porosity [31]. Gu et al. investigated void formation by transforming hygroscopic water absorbed by glass and carbon fibers, as well as trapped air, into vapor bubbles as a result of the temperature rise during the thermocompression process [6]. Compared to synthetic fibers, the problems associated with the processing of plant fiber composites are more complicated due to the particular characteristics of this type of fiber. The use of plant fibers for resin reinforcement necessitates careful consideration of the production conditions. The essential criteria for regulating the thermal deterioration of the fibers are the process temperature and time. To reduce viscosity, the hot-molding temperature must be higher than the melting point of the resin, and the time must be long enough to allow the molten resin to permeate the fibers, assuring good adhesion between the reinforcement and the matrix. In conflict with these needs, the melting temperature and time should be as low as possible to slow down the thermal deformations that occur and cause fissures and permanent damage to the fibers, as well as the pectin breakdown, which begins at 180°C [32, 33].

5. Applications

E-glass fibers are commonly used as reinforcements in shipbuilding, while carbon fibers saturated with epoxy resin are commonly used in aeronautics. The use of synthetic fibers in composites is supported by their high chemical resistance, compatibility with most impregnation resins, and mechanical and thermal performance. However, the usage of this sort of reinforcement is no longer adequate: On the one hand, their comparatively large density penalizes them; on the other hand, they endanger the health and the environment. Since the 1980s, these environmental problems have become a major concern for our society and the media. And since then, the industrial optimization of eco-composites is booming thanks to their high specific mechanical properties. The limits of applications are constantly being pushed back through the development of fiber preforms and the adaptation of processing methods.

5.1 Automobile

Automobile body sections, such as engine hoods, dashboards, and storage tanks, are made using natural fiber reinforcements such as flax, hemp, jute, sisal, and ramie. The VARTM manufacturing technology was used for these composite constructions, and its liability was tested through structural testing and impact stress analysis. As a consequence, the material's weight was reduced while its stability and strength were improved. The increase in safety characteristics was tested using the head impact criterion (HIC), and it was discovered that composite constructions with natural fiber reinforcements are appropriate for automotive body sections [34–36]. **Figure 5** depicts the external body elements of a Volkswagen x11 crazy carbon fiber replica.

The automobile sector, in particular, has shown a genuine commitment to economic and environmental concerns by using natural fibers in different non-structural components (dashboards, door panels, spare wheel covers, etc.) with the goals of lowering mass, fuel consumption, and emissions (**Figure 5**).

5.2 Aerospace

Fiber-reinforced epoxy composites manifest the properties required for aircraft interior panels, such as resistance to heat and flame and disposal of materials. Fiber-reinforced epoxy composite shows a variety of applications in the aerospace industry

Figure 5.
Volkswagen xl1 carbon fiber body pieces, adapted from [36] under a creative commons license. (a) the 45 kg of natural fibers in a Mercedes S-class. (b) Car door panel made of natural fibers. (c) Spare wheel cover made of natural fibers.

due to its superior mechanical properties and lightweight structure. Conductive fibers in the layer of fiber composite structure eliminate the requirement of separate wires for transceivers of communication devices. High stiffness with a lower coefficient of thermal expansion is achieved when P100 graphite fibers diffused in 6061 aluminum matrix composite material are employed to the high gain antenna of the Hubble space telescope [37].

For example, the wing of the plane is a composite material, the fiber is carbon fiber and the resin is epoxy. The manufacturing technique involves resin infusion: all the reinforcing fibers are dried, shaped and then the resin is infused into the reinforcement. The choice of polymer matrix must both ensure good performance for the finished wing after curing and also maintain a well-tuned reactivity, not too high to allow the wing to be infused, which can take several hours, but enough to allow the reactions to take place effectively.

5.3 Marine

Components and structures functioning in the marine environment are subjected to significant stresses caused by wind, waves, and tides. Furthermore, they must endure hostile and harsh environmental conditions throughout their lives, including being placed in the splash zone if not submerged in seawater. The use of polymer composites in maritime systems has been the subject of much research in recent decades, showing the potential benefits of replacing various components such as ship hulls, propeller blades, wind, and tidal turbine blades, to name a few [38]. For example, in the offshore construction (seawater piping, stairways and walkways, firewater piping, grating, fire and blast walls, cables and ropes, storage vessels, and so on), valves and strainers, fans and blowers, propeller vanes, gear cases, condenser shells, and so on.

And more other applications, which we will not be able to represent all of them such as:

- Civil engineering includes the construction of new advanced structures (roofs, plate and shell elements, linear elements, pipes and tanks, folded structures, and so on) as well as the rehabilitation of existing metallic and concrete structures such as buildings, bridges, pipelines, masonry construction, and so on.
- Sporting goods: Golf club shafts, tennis rackets, bicycle frames, fishing rods, and so forth.
- Electrical and electronic components include power line insulators, fiber optics tensile members, lighting poles, and so on.
- Chemical Industries: stacked bottles for fire departments, composite containers for substances, mountain climbing, ducts and stacks, subterranean storage tanks, and so on.
- Medical field: Orthopedic medicine, prosthetic devices, and imaging (MRI).

6. Challenges

Understanding the significant material characteristics of fiber/epoxy constituents, as well as the fundamental structures and availability of production technologies, is required for the use of fiber/epoxy composites in a range of applications.

Furthermore, the manufacturing technique used has an effect on the ultimate qualities of the material. The cost of materials is influenced by production volume—the bigger the volume of production, the lower the cost of materials. In the instance of the car industry, increasing production volume increases the risk of investing in raw materials while building manufacturing set-up based on production rate and cycle time. In addition, the product's design complexity increases the cycle time, decreasing the manufacturing pace.

7. Conclusions

On fiber-reinforced composite manufacturing, current progress, novel advancements, and future research prospects are summarized and presented. However, the ongoing demand for composite constructions necessitates a large consumption of environmentally hazardous components. Certain fibers (for example, carbon fibers) utilized to improve qualities in numerous sectors are a significant hindrance to recycling at the end of the composites' life. As a result, the current environmental crisis, which has reached a tipping point, necessitates immediate and objective action to cut greenhouse gas emissions. As a result, obtaining advanced composites from renewable energy resources would be the best ecological answer. Furthermore, future research areas might focus on recycling current composites into high-value alternative goods. Furthermore, new innovative methods for post-consumer waste treatment must be developed. Additionally, new sophisticated technologies for post-consumer waste treatment must be developed, or existing FRP composite production technologies must be improved.

Conflict of interest

The authors declare no conflict of interest.

Author details

Nassima Radouane[1,2]* and Abdelkrim Maaroufi[2]

1 UDSMM (EA 4476), MREI-1, Université du Littoral Côte d'Opale, Dunkerque, France

2 Laboratory of Composite Materials, Polymers and Environment, Department of Chemistry, Faculty of Sciences, University of Mohammed V, Rabat, Morocco

*Address all correspondence to: nassima.radouane@univ-littoral.fr

References

[1] Du Y, Li D, Liu L, Gai G. Recent achievements of self-healing graphene/polymer composites. Polymers. 2018 Feb;10(2):114

[2] Linul E, Vălean C, Linul PA. Compressive behavior of Aluminum microfibers reinforced semi-rigid polyurethane foams. Polymer. 2018;**10**:1298. DOI: 10.3390/POLYM10121298

[3] Chukov D, Nematulloev S, Torokhov V, Stepashkin A, Sherif G, Tcherdyntsev V. Effect of carbon fiber surface modification on their interfacial interaction with polysulfone. Results in Physics. 2019;**15**:102634. DOI: 10.1016/J.RINP.2019.102634

[4] Ortego A, Valero A, Valero A, Restrepo E. Vehicles and critical raw materials: A sustainability assessment using thermodynamic rarity. Journal of Industrial Ecology. 2018;**22**:1005-1015. DOI: 10.1111/jiec.12737

[5] Velandia Vargas JE, Falco DG, da Silva Walter AC, Cavaliero CKN, Seabra JEA. Life cycle assessment of electric vehicles and buses in Brazil: effects of local manufacturing, mass reduction, and energy consumption evolution. International Journal of Life Cycle Assessment. 2019;**24**:1878-1897. DOI: 10.1007/S11367-019-01615-9

[6] Jagadeesh D, Kanny K, Prashantha K. A review on research and development of green composites from plant protein-based polymers. Polymer Composites. 2017;38:1504-1518

[7] Myers RR. Epoxy resins: Chemistry and technology. Journal of Colloid and Interface Science. 1974;49:162

[8] Feng Z, Adolfsson KH, Xu Y, Fang H, Hakkarainen M, Wu M. Carbon dot/polymer nanocomposites: From green synthesis to energy, environmental and biomedical applications. Sustainable Materials and Technologies. 2021;**29**:e00304. DOI: 10.1016/J.SUSMAT.2021.E00304

[9] Gan, S.; Shahabudin, N.; 2019, Applications of Microcapsules in Self-Healing Polymeric Materials.

[10] Suliga A, Hamerton I, Viquerat A. Cycloaliphatic epoxy-based hybrid nanocomposites reinforced with POSS or nanosilica for improved environmental stability in low earth orbit. Composites Part B: Engineering. 2018;**138**:66-76

[11] Chekanov Y, Arrington D, Brust G, Pojman JA. Frontal curing of epoxy resins: Comparison of mechanical and thermal properties to batch-cured materials. Journal of Applied Polymer Science. 1997;**66**:1209-1216. DOI: 10.1002/(SICI)1097-4628(19971107)66:6

[12] Vieille B, Casado VM, Bouvet C. About the impact behavior of woven-ply carbon fiber-reinforced thermoplastic- and thermosetting-composites: A comparative study. Composite Structures. 2013;**101**:9-21. DOI: 10.1016/J.COMPSTRUCT.2013.01.025

[13] Bouvet C, Rivallant S, Barrau J-J. Low velocity impact modeling in composite laminates capturing permanent indentation. Composites Science and Technology. 2012;**72**(16):1977-1988

[14] Nemati Giv A, Ayatollahi MR, Ghaffari SH, da Silva LF. Effect of reinforcements at different scales on mechanical properties of epoxy adhesives and adhesive joints: A review. The Journal of Adhesion. 2018 Nov 10;94(13):1082-1121

[15] Patnaik A, Satapathy A, Chand N, Barkoula NM, Biswas S. Solid particle erosion wear characteristics of fiber and particulate filled polymer composites: A review. Wear. 2010;268:249-263

[16] Shoaib M, Shehzad A, Omar M, Rakha A, Raza H, Sharif HR, et al. Inulin: Properties, health benefits and food applications. Carbohydrate Polymers. 2016;**147**:444-454. DOI: 10.1016/j.carbpol.2016.04.020

[17] Radouane N, Hammi M, Maaroufi A, Ouaki B, Depriester M, Hadj-Sahraoui A. Investigation on wettability and mechanical properties of novel zinc phosphate glass-based epoxy composites. In: Proceedings of the IOP Conference Series: Earth and Environmental Science. Vol. 690. China: IOP Publishing Ltd; 2021. p. 12047

[18] Prakash S. Experimental investigation of surface defects in low-power CO2 laser engraving of glass fiber-reinforced polymer composite. Polymer Composites. 2019;**40**:4704-4715. DOI: 10.1002/PC.25339

[19] Chalmers DW. Experience in design and production of FRP marine structures. Marine Structures. 1991;**4**(2):93-115

[20] Rajak DK, Pagar DD, Menezes PL, Linul E. Fiber-reinforced polymer Composites: Manufacturing, properties, and applications. Polymer. 2019;**11**:1667. DOI: 10.3390/POLYM11101667

[21] Devendra K, Rangaswamy T. Strength characterization of E-glass fiber reinforced epoxy Composites with filler materials. Journal of Minerals and Materials Characterization and Engineering. 2013;1:353-357. DOI: 10.4236/jmmce.2013.16054

[22] Kumar P. An experimental and numerical investigation of mechanical properties of glass Fiber reinforced epoxy Composites supercritical natural circulation loop view project. Advanced Materials Letters. 2013;**4**(7):567-572. DOI: 10.5185/amlett.2012.11475

[23] Khan ZI, Arsad A, Mohamad Z, Habib U, Zaini MAA. Comparative study on the enhancement of thermo-mechanical properties of carbon fiber and glass fiber reinforced epoxy composites. Materials Today: Proceedings. 2021;**39**:956-958. DOI: 10.1016/J.MATPR.2020.04.223

[24] Nur Azrie Bt Safri S, Sultan MTH, Jawaid M. Damage analysis of glass fiber reinforced composites. Durability and Life Prediction in Biocomposites, Fibre-Reinforced Composites and Hybrid Composites. 2019:133-147. DOI: 10.1016/B978-0-08-102290-0.00007-6

[25] Ayyanar CB, Mohan SKP, Bharathiraj C, Mavinkere Rangappa S, Siengchin S. Characterization of Syzygium cumini particulates filled E-glass fiber-reinforced epoxy composites. Polymer Composites. 2021;**42**:6298-6309. DOI: 10.1002/PC.26298

[26] Kumre A, Rana RS, Purohit R. A review on mechanical property of sisal glass fiber reinforced polymer composites. Materials Today: Proceedings. 2017;**4**:3466-3476. DOI: 10.1016/J.MATPR.2017.02.236

[27] Radouane N, Depriester M, Maaroufi A, Singh DP, Ouaki B, Duponchel B, et al. Synthesis, mechanical, thermal, and electrical characterization of graphite–epoxy composites. Journal of the Chinese Chemical Society. 2021;**68**(8):1456-1465. DOI: 10.1002/jccs.202000490

[28] Gantayat S, Rout D, Swain SK. Carbon nanomaterial–reinforced epoxy

Composites: A review. Polymer-Plastics Technology and Engineering. 2018;**57**: 1-16. DOI: 10.1080/03602559.2017.1298802

[29] Xu W, Ke G, Wu J, Wang X. Modification of wool fiber using steam explosion. European Polymer Journal. 2006;**42**(9):2168-2173

[30] Hou W, Zhang W. Advanced composite materials defects/damages and health monitoring. In: Proc. IEEE 2012 Progn. Syst. Heal. Manag. Conf. PHM-2012. 2012. DOI: 10.1109/PHM.2012.6228804

[31] Liu L, Zhang B, Wang D, Wu AJ. Effects of cure cycles on void content and mechanical properties of composite laminates. Composite Structures. 2006;**73**(3):303-309

[32] Nechyporchuk O, Belgacem M, Bras J. Production of cellulose nanofibrils: A review of recent advances. Industrial Crops and Products. 2016;**93**:2-25

[33] Khanlou H, Woodfield P, Summerscales J, Hall W. Consolidation process boundaries of the degradation of mechanical properties in compression moulding of natural-fibre bio-polymer composites. Polymer Degradation and Stability. 2017;**138**:115-125

[34] Shojaeefard MH, Najibi A, Rahmati Ahmadabadi M. Pedestrian safety investigation of the new inner structure of the hood to mitigate the impact injury of the head. Thin-Walled Structures. 2014;**77**:77-85. DOI: 10.1016/j.tws.2013.11.003

[35] Kong C, Lee H, Park H. Design and manufacturing of automobile hood using natural composite structure. Composites. Part B, Engineering. 2016;**91**:18-26. DOI: 10.1016/j.compositesb.2015.12.033

[36] Belauto. 2014-volkswagen-xl1-carbon-fiber-body-parts.jpg (1500×938). Available on: https://belauto.com.my/wp-content/uploads/2014/12/2014-volkswagen-xl1-carbon-fiber-body-parts.jpg [Accessed: 28 02 2022]

[37] Jom SR. Metal-matrix composites for space applications. The Journal of The Minerals, Metals & Materials Society. 2001;**53**:14-17. DOI: 10.1007/s11837-001-0139-z

[38] Rubino F, Nisticò A, Tucci F, Carlone P. Marine application of fiber reinforced Composites: A review. Journal of Marine Science and Engineering. 2020;**8**:26. DOI: 10.3390/JMSE8010026

Chapter 4

Synthesis and Properties of Epoxy-Based Composites

Srikanta Moharana and Bibhuti B. Sahu

Abstract

Epoxy-based composites are of great interest among academic and industrial researchers owing to their low cost, superior mechanical properties, large specific strength, super adhesiveness with good thermal and solvent resistance in recent times. However, the effect of carbon-based nanofiller reinforced epoxy composites is of prime focus due to their significant mechanical, dielectric and electrical performances for technological applications in broad fields of nanoscience and technology. There is a greater influence on the properties of the nanofiller reinforced epoxy matrix composites depending on the concentration of various types of nanofillers. The processing techniques play a crucial role in the prediction of attractive and suitable properties of the various nanofiller reinforced epoxy composites. There are several processing methods that have been employed to accomplish a superior degree of dispersion of nanofillers in the epoxy matrix. This current chapter portrays the simultaneous focus on their preparation techniques and effect of the dielectric, electrical and mechanical properties of various carbon nanofillers (such as fullerene, carbon nanotubes (CNTs), carbon nanofibers (CNFs) & graphene) filled epoxy resin composites for a broad spectrum of technological applications. We hope this chapter will facilitate the concrete in-depth ideas to the readers on the progress of various synthesis techniques and properties of different nanofiller reinforced epoxy composite systems.

Keywords: epoxy, nanofiller, carbon nanotubes, graphene, composites, mechanical, dielectric, electrical, properties

1. Introduction

The emergence of new technological fields is associated with the development of new hybrid polymeric composite materials with high-performance practical applications. These composite materials have several interesting multifunctional properties including superior strength, high stiffness or modulus of elasticity, durability, corrosion resistance, better thermal stability, enhanced electrical and electronic properties, lightweight with highest specific stiffness and strength along the direction of the reinforcing fiber, dimensional stability, good temperature, chemical resistance, flex performance and ease of processing with cost-effectiveness in contrast to other types of material [1–3]. However, there are different composite systems, which are used in the field of manufacturing technology. Among, epoxides, phenolics, polyurethane and polyimides are commonly used as the matrix of the materials for the progress of advanced hybrid composite materials. The epoxy resin is one of the significant polymeric materials among academic and industrial researchers due to its remarkable versatile properties like thermal, electrical, mechanical performances, superior thermal and chemical resistance [4–7]. The thermoset resin such as epoxy is currently in wide use for the composite industry. This is because of their significant chemical, corrosion resistance, excellent adhesion performance, low shrinkage and lesser price with their challenging applications. Generally, the cured epoxy gives rise to large modulus, strength, good resistance to creep and high performance at elevated temperature due to its extreme cross-linked microstructure. The improved toughness of cured epoxy resin is commonly advantageous owing to its poor ductility. The betterment of toughness is a proficient approach for incorporating rigid or reactive rubbery particles into the epoxy network. It is the consequence at the expense of glass transition temperature and strength of this polymeric matrix [8–10]. Epoxy is intensely associated with daily life in the form of packaging, coating, adhesive and electrical insulating materials as well as applications in electrical appliances, semi-conductors, etc. [10, 11].

In the descended decade's, aluminum is one of the common metals around the globe, which is about 8% of abundance on earth's crust and seen in the form of oxides including karolinite, bauxite, nepheline and alunite. Sir Humprey Davy in the year 1808 revealed the existence of aluminum and further Oersted in 1825 formed its tiny pellets. However, Wohler a German scientist (1845) verified the specific gravity and aluminum lightness. They also discovered certain performances such as ease of deformation, air stability and its melting with a blow torch. The metals are characterized by high corrosion resistance, superb machining performances, superior thermal and electrical properties, large ductility, low strength, hardness and wear resistance [12, 13]. Due to high corrosion resistance and its lightness, aluminum and its alloy based materials played a vital role for the production of equipment (panels, roofs and frames) of packaging materials in the area of food and transportation (vehicle and aircraft parts) [13, 14]. Due to the broad spectrum of attractive properties and potential applicability of aluminum academic and industrial researchers have converged their focus on increasing the strength of aluminum and its hardness through solid solution and hardening. Also, they have progressed the aluminum-based metal matrix composite materials by the reinforcement of various fillers in the matrix of aluminum [12–14]. Recently, various researchers are putting their efforts to attempt for developing appropriate materials using aluminum-based alloys for end used applicability in the field of aerospace application, cast aluminum engine is used on flier, the manufacture of wind ribs (aluminum 2050). The high static strength of aluminum 7079 and 7075 are used to give sufficient toughness and corrosion resistances [3, 15, 16]. The composite of aluminum-graphite shows superior thermal conductivity as a result of the appreciable contribution coming out of the metal matrix. The aluminum-based metal matrix composite systems preserve the advantageous properties of both the reinforcement and matrix by associating the vital strength of the reinforcement with the ductility of the aluminum matrix [17, 18]. Many researchers have reported about the synthesis of carbon black powders by using agricultural byproducts including coconut shell, apricot stones, sugarcane, bagasse, nutshells, tobacco stems and forest residues. It is observed that the coconut shell shows costly disposal and cause environmental problem. Thus, they adopted an appropriate technique for the synthesis of carbon black using coconut shell owing to their superior natural structure and small ash contents *via* pyrolysis route (carbonization of coconut shell) with the application of temperature ranging from 550°C to 900°C. [3, 19–22]. However, activated carbon developed through the conversion of coconut shell can be used as filler in processing the composites, which have potential utility in significant adsorbent for purification of water or industrial treatment and municipal effluents. The addition of these filler may also diminish the cost for waste disposal with a cheap alternative as compared to the commercial carbons [3, 21–24]. The byproducts (barley husk and coconut shell) reinforced thermoplastic is a better alternative for wood fibre-based hybrid composite materials. The experimental results revealed that barley husk and coconut shell are thermally stable at high temperature than that of soft wood fiber with different percentages of cellulose content (50% barley husk and 34% coconut shell) [25, 26]. Both coconut shell and barley husk are of large carbon-rich layers on their surface as compared to soft wood fiber. The superior tensile strength is observed in the barley husk fiber reinforced composite than that of the soft wood fibre-based composite. Moreover, the coconut shell and barley husk reinforced composite exhibited 80% and 40% improved elongation at break, 20 and 35% superior impact strength as compared to soft wood fiber composite systems [22, 27, 28].

The synthesis of composites is the combination of two materials such as matrix and reinforcement to form a hybrid composite material with excellent electrical and mechanical performances. The matrix is in the bulk form, which employs reinforcement with a strong bond [29–31]. However, the reinforcement is normally embedding additives for enhancing the properties of the material. The thermoplastic and thermosetting polymers are usually used as a matrix in the polymer composites. On the other hand, thermosetting polymer matrix exhibits better stiffness and superior strength than that of the thermoplastic polymeric materials [32]. Epoxy resin is one of the most commonly used thermosetting polymer matrix with better mechanical performances and good adhesive property with the incorporation of reinforcement particles [33, 34]. Thus, epoxy resin reinforced filler-based composites have immense interest among scientist, researchers for the development of hybrid composites for their use in the field of composite manufacturing industries, automobiles, paints and coating industries [35, 36]. The epoxy-based materials have common shortcomings including low impact strength and weak wear resistance, which can be overcome by selecting proper reinforcement of filler particles in the matrix; this will be useful in the field of tribological application. In epoxy-based composites, the epoxy reinforced nanoparticles show the large surface area with substantial interaction between matrix and filler particles [35–39]. However, nanocomposites are generally lightweight than that of the micro composites due to the relatively high density of the micro-additive fillers [40]. The enhancement of wear resistance and mechanical performances is due to the incorporation of hard oxide and carbide-based nanoparticles (including silicon oxide, alumina and tungsten) into the matrix [41–43]. A significant technique for the preparation of nanocomposites is needed for the homogeneous dispersion of the nanoparticles reinforced epoxy matrix *via* sonication technique [44]. The most crucial factors which influence the properties of the composites are curing conditions, molecular bonding between reinforcement and epoxy matrix with ratio of curing agent [45, 46]. There are various researchers have made to synthesize epoxy resin reinforced nano-filler-based hybrid composites for improving the mechanical and electrical performance of the composite systems. The different nanofillers (such as alumina, fullerene, graphene, carbon nanotubes (CNTs), carbon fibers (CFs), etc.) reinforced epoxy matrix composites result superior thermal conductivity, large thermal stability and better wearable resistance [44–47].

This chapter is organized on the basis of various filler-based nanomaterials used in epoxy resin composites, which includes carbon-based nanomaterials, fullerene, graphene, CNTs, CFs, nanoclay reinforced epoxy resin composites have been presented in minute details in Section I. This section comprises of a brief introduction of the specific filler materials and its effect on epoxy resin followed by detailed discussion on the filler reinforced epoxy composite systems. Section II portrays the vivid insight into the synthesis techniques of carbon-based nano-filler reinforced epoxy composites. In Section III, we have especially emphasized on dielectric and mechanical properties of different filler reinforced epoxy resin composite systems.

2. Different carbon based nanofillers reinforced epoxy composites

Carbon-based epoxy composites are considered to be one of the most promising groups among the advanced materials of current times due to their distinctive physical and chemical properties. These materials exhibit significant properties as a result of the introduction of nanofillers into the matrix materials gives rise to unexpected properties, which make them distinguished due to unparallel design possibilities. In this context, Kroto et al. [48] in 1985 have discovered fullerene, which is the allotrope of carbon, where in fullerene the molecules of carbon atoms are well connected by single and double bonds. The family of fullerene is then extended with other forms of synthesis of carbon-based nanomaterials like single-walled carbon nanotubes (SWCNTS) (1991), multiwalled carbon nanotubes (MWCNTs) (1993) and graphene

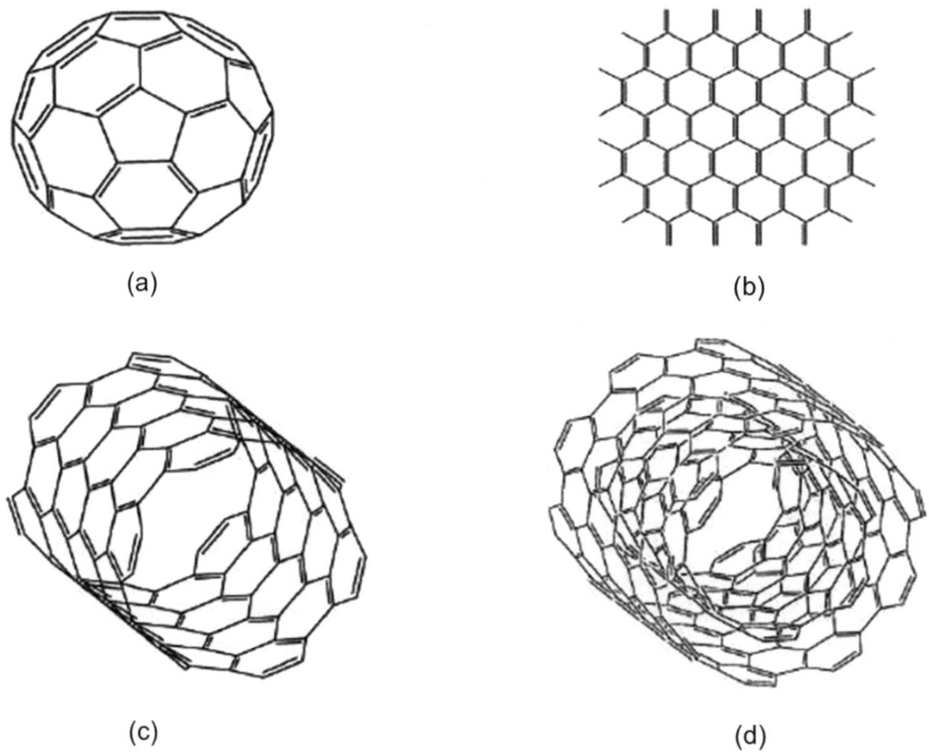

Figure 1.
Carbon-based nanoparticles (a) graphene, (b) fullerene, (c) SWCNTs and (d) MWCNTs. Reprinted with permission from Ref. [49].

(2004) (as shown in **Figure 1**) have gained immense attention to the scientists for further research [49–55].

The carbon-based graphene nanofillers [49, 53–55] play an important role for the synthesis of composites in the field of nanoscience and technology. A single layer carbon sheet of graphene with sp^2 hybridization (two dimensional) is arranged in a hexagonally packed lattice structure analogous to a honey comb. They have unique performances including high charge mobility at room temperature, high surface area, good optical transparency, large young's modulus and superior conductivity. However, it is observed that fullerene may be considered as fascinating reinforcements in comparison to CNT or graphene due to their zero-dimensional structure of carbon molecules, which exist in the form of spherical, tube shape and ellipsoid. The fullerene related to C_{60} is called Bucky ball or Buckminster fullerene spherical in shape, which corresponds to a soccer ball [49, 54–56]. The carbon nanotubes (CNTs) [57–61] a new allotrope of carbon is normally thin hollow cylindrical fullerene structure in the diameter of nano-scale range and little micron length with significant properties. Recently, CNTs are considered to be the most talented candidates as reinforcement for polymer composites due to their high aspect ratio, high young's modulus, tensile strength, large thermal conductivity and approximately 1000 times larger than that of the electrical capacity than copper and thermally stable in vacuum at 2800°C [60, 61].

According to the number of concentric layers of carbon atoms, CNT is available at single-walled carbon nanotubes (SWCNTs) and multiwalled carbon nanotubes (MWCNTs), which shows multiple SWCNTs arranged in a concentric and coaxial manner as well as more flexible in nature than that of the MWCNTs. Moreover, CNTs appeared as outstanding materials and can be used as conducting nanofillers in the polymer matrix to give high-performance composites. Similarly, one dimensional carbon nanofibers (CNFs) [8] show a hollow cylindrical structure with lower cost and ease of processing than that of the pure CNTs. However, the carbon-based materials with a diameter less than 500 nm, do not exhibit structure of CNT falls into the class of CNFs and CNT, which exhibits clear structure as compared to CNFs. In this section, there are various studies have been made to achieve significant properties in epoxy-based carbon nanofillers composites [8, 57–61]. For instance, Kim et al. [62] have reported surface modified epoxy-based CNTs composites with superior homogeneity of CNTs in the epoxy matrix. They also studied their effect on the rheological and mechanical performances of the resultant composites. The multiwalled carbon nanotubes (MWCNTs) reinforced epoxy resin composites are of uniform dispersion in the matrix through the ultrasonication technique. The synthesized MWCNTs reinforced epoxy composites have superior Young's modulus and strength with optimized parameters were reported by Montazeri and his co-workers [63]. Allaoui et al. [64] have fabricated MWCNTs based rubber epoxy matrix composites with different concentrations of MWCNTs contents. It is observed that the resultant composite system has enhanced electrical performance with the increase of filler concentration in the epoxy matrix. However, the lower weight percentage (0.5 wt%) of MWCNTs reinforced epoxy composites exhibits a considerable increase of tensile strength and Young's modulus were reported by Montazeri and his group [65]. The incorporation of MWCNTs into the epoxy composites showed significant improvement of the thermal and mechanical performances *via* oxidation, acylation and amidation. It is also noticed that the enhancement of the electrical conductivity of the silane modified MWCNT-epoxy composites as compared to unmodified MWCNT-epoxy composites has been reported by Shen et al. [66]. Similarly, Choi et al. [67] have studied the effect of silver-plated MWCNTs into the epoxy matrix. It is also noticed that the synthesized composites are of improved thermal conductivity with the increase of filler contents and time duration.

2.1 Fullerene reinforced epoxy composites

The contemporary demand for the development of lightweight hybrid composite materials with high specific strength, stiffness and improved tribological performance for end used applicability in the areas of aerospace and automotive industries [49, 56, 68]. During the past few times, fullerene and fullerene-based composites have been synthesized and used in various applications particularly in the thin films, organic polymers and hybrid organic-inorganic composites in the field of microelectronics [68]. There are various established techniques for the preparation of fullerene-based hybrid composites including solid-state reaction technique, liquid state techniques, deposition and spraying technique [56]. Rafiee et al. [69] have fabricated fullerene reinforced epoxy composites with various concentrations of filler content. It is revealed that the resultant composites with Young's modulus, fracture toughness, ultimate tensile strength were significantly improved in the epoxy matrix. It is reported that the incorporation of fullerene into the epoxy matrix shows superior properties as compared to neat nanosilica, nano-alumina and nano-titania filled epoxy composites; this may be due to the hollow structure, which results in increasing surface area and reduced weight simultaneously. However, the dispersion of fullerene into polymer matrix is easier than that of the other carbon based nanofillers which includes CNTs and CNFs, etc. The 1D and 2D filler-based materials are more prone to entanglement to each other than spherical fullerene particles. There are various research works related to fullerene modified polymer composites were reported by Ayesha et al. [70, 71]. However, the reports have been made based on fullerene modified epoxy-fiber composites. For example, fullerene modified epoxy carbon fiber reinforced polymer composites were fabricated and studied their effect of various mechanical performance of fullerene concentrations in the matrix. It is observed that 0.5% of epoxy matrix composites have improved interlaminar fracture toughness (60%) and also enhanced tension and compression up to 12% were reported by Ogasawara et al. [72]. Similarly, Jiang et al. [73] have reported about fullerene modified epoxy (1–3% concentration) reinforced fiber composites show improvement in the bonding strength between unidirectional carbon fiber and matrix. This improvement may be due to the suppression of fracture at interfacial layers of fibers, which is attributed to the presence of fullerene nanoparticles in the polymer matrix.

2.2 Carbon nanotube and carbon nanofiber reinforced epoxy composites

Several techniques including chemical vapor deposition (CVD), arc discharge and laser ablation techniques have been used to synthesize carbon nanotubes and CNT reinforced polymer composites in the past few decades [74–84]. The improvement of physical properties of the epoxy composites by the incorporation of different carbon-based nanofillers was reported by Liu et al. [85]. One of the key problems for the preparation of carbon nanotubes-epoxy composite system is the agglomeration or aggregation in the matrix. Thus, various techniques and studies related to CNT-based epoxy composites have been adopted to enhance the better dispersion and reduction of agglomeration or aggregation in the matrix [86, 87]. Moreover, the direct use of CNT in the epoxy matrix without surface treatment may give a marginal improvement in the composites. Besides, the unmodified CNT embedded into the epoxy resin with one dimensional structure forms poor bonding, which is the high tendency to give entanglement with each other results in some problems. The structural performance is enhanced with the application of strong acids to convert the C–C bonds into various functional groups of amine, amide, etc. on the side surface of the CNT *via* plasma treatment or UV/ozone treatments [88]. In addition, the attachment of CNTs into the metal particles with enhanced homogeneity and interfacial adhesion results for improving the structural properties as compared to neat CNT-epoxy composites [89]. Similarly, carbon nanofibers are also creating some difficulties in uniform dispersion of these particles in the epoxy matrix owing to their structure. There are various physical (sonication, mechanical stirring, plasma treatment, high-temperature heat treatment) and chemical (surface functionalization through modifying agents, surfactants, etc.) methods are employed to synthesize homogeneous dispersion of CNF in the epoxy matrix. It is also observed that, using these techniques for preparation may affect the properties of the bulk composites. For example, the high-temperature heating technique for preparation of composites may reduce the interfacial strength, this lead to decrease the structural properties [90–93]. The high concentration of filler in the polymeric matrix may affect the final properties of the composites. To overcome this shortcoming, the incorporation of lower filler loading into the matrix gives better homogeneity as well as enhance structural properties of the composite systems. Moreover, for the utilities of CNTs various optimized parameters are essential to achieve better homogeneity, desired orientation and functionalization to extract the utmost benefits of this marvel material.

2.3 Graphene reinforced epoxy composites

The two-dimensional macromolecule graphene and its derivative have extensively been explored due to their application as nanofillers in graphene reinforced composites. The production of nanofillers must be attainable at a large scale with low cost for the progress of graphitic fillers from natural graphite, which make them suitable nano-filler-based materials for reinforcement in the field of composites *via* simple processing technique. These graphene-based composites are of immense interest among academic and industrial researchers owing to the excellent electrical, thermal and mechanical performance of graphene [49, 54, 55, 94–97]. However, there are various nanofillers including graphene nano platelets or sheets (GNPs or GNs) and graphene oxide (GO) shows additional flexibility at the nano-scale due to the stability of macromolecule and multitude of alternatives for more functionalization for the composite systems. An enormous modified forms of carbon fiber and polymer phase of carbon fiber reinforced composites (CFRC) are found in various literatures. The ultrafine GNP has been utilized to exfoliate graphite flake producing graphite nanoplatelets (GNP), on the other hand graphene oxide (GO) *via* modified hummer's technique exhibits oxidation followed by exfoliation of bulk graphite [96, 97]. Graphite is of strong planar structure with each carbon atom forming three covalent bonds with adjacent atoms. Whereas graphene oxide (GO) contains different functional groups (hydroxyl, epoxy and carboxylic acid groups) joined through sp^3 hybridized carbon atom but it may partially retain a number of sp^2 hybridized carbon depending on their reaction conditions. Moreover, it is also observed that the presence of polar functional group present in the surface of the graphene oxide, which results in bonding with polymer matrix to form strong interfaces. It is reported that the significant properties of graphene play an important role in the scientific research communities to develop graphene-based epoxy matrix for the application in structural composites. Besides, these fillers are applied for additional needs to develop composites with attractive properties like thermal, electrical conductivity and mechanical strength [98–101]. Alexopoulo et al. [102] reported on the fabrication of epoxy-based GNP composite systems and studied their various properties on the effect of size on the GNP particle embedded into the epoxy matrix. It has also been noticed that the synthesized composites show the formation of agglomeration at higher concentrations (>5 wt%) within the matrix. Further, at lower concentration (0.25 wt%) of GNs, it is revealed that there is considerable improvement in properties (toughness, flexural strength and flexural modulus) of the composites.

The epoxy-based composites using reduced graphene oxide act as fillers, which give rise to significant improvement in properties (tensile, impact and flexural) of the resultant composite systems. However, it is interesting to observe that the functionalised graphene nano-sheet (GNS) reinforced epoxy composite at lower filler concentration shows superior fracture toughness, fracture energy, stiffness, strength, and fatigue resistance than that of the neat CNT based epoxy composites [103–106]. The advantages of GNS as compared to CNTs raise the structural properties of epoxy, which may be attributed to the larger surface area with improved adhesion of filler-matrix owing to the wrinkled surface and the two-dimensional geometry of GNS [105, 106]. Graphite continues to attract considerable attention among researchers due to its excellent mechanical, electrical properties, low density with ease of processing and low cost. However, graphite commonly exists as a layered material and these layers are closely packed through Van der Waals force. For the efficient use of graphite as filler in polymer composites, its layer must be separated partly to achieve expanded graphite (EG). It is difficult to intercalate monomers into the interlayer of graphite to produce composites if the raw graphite is used as reinforcement; also, it is not possible to disperse graphite layers in the epoxy matrix. To overcome this adversity, the preparation of expanded graphite (EG) from raw graphite is exposed to a strong oxidizer (e.g., HNO_3, H_2SO_4, $KMNO_4$). The EG and GO both are the derivatives of graphite and are considered to be ultimate nano-filler materials for epoxy matrix. It is due to the presence of covalent bonds having hydroxyl, phenolics and epoxide functional groups on their basal planes and also it is located on the carboxyl and carbonyl group at the sheet edges. The presence of these functional groups enables them to strongly hydrophilic in nature. The EG is readily dispersed in water and reinforced into matrix with the help of these functional groups for the synthesis of composites [107–109]. However, two-dimensional graphene-based nanomaterials show little agglomeration because of their high aspect ratio, which affects the mechanical properties of the resultant composites. Therefore, suitable dispersion and exfoliation techniques are essential to facilitate better structural properties.

2.4 Nanoclay reinforced composites

The nano-clay reinforced polymer-based composites have earned much attention among both academic as well as industrial sectors due to incorporation in a small amount of nanoclay considerably improves the mechanical performance of the neat polymers. The nanoclay based two dimensional nano-material are naturally occurring in the form of platelets, which include a few to 1000 sheets. These are mainly silicate and comparatively inexpensive than that of the other nanomaterials. The other types of nanomaterial possess a larger surface area with a high aspect ratio (>50) and are thermally stable. Usually, the use of nanoclay in the matrix improves their properties with good optical transparency for a suitable selection of nano-filler reinforced in the polymer composites [49, 110].

Montmorillonite (MMT) is an aluminosilicate [111], which is extensively used in the clay based nanofillers (**Figure 2**). The thickness and lateral dimensions are observed in the montmorillonite around 5 nm and 500 nm, respectively. The single sheet of montmorillonite has a stiffness of about 250 GPa. Moreover, the large stiffness was observed in the clay minerals, which makes them appropriate for improving the structural performance of the polymer-based composites [111–113]. In recent few times, it has been reported about the epoxy-based glass and carbon composites, which may be embedded in the glass reinforced polymer composites. In addition, a montmorillonite (Nanomer I.30E) based clay mineral was used in this research [49]. They have reported each platelet are approximately 1 nm thickness and internal dimension around 300–600 nm with a high aspect ratio. Moreover, the compression strength is improved about 15–20% for particularly 5 wt% of nanoclay reinforced epoxy matrix composite systems. It is also observed that the incorporation of a small amount of nanoclay into the epoxy composites enhanced the mechanical performance using reinforcement of clay particles in the polymer matrix, which results the increase in impact and interfacial properties of the composite systems. Especially, the current research is based on nanoclay composites, which consist of the polymer as matrix material and nanoclay as the reinforcement particle (act as nanofiller) in the composite systems. Several properties (including mechanical, thermal and electrical) and structural aspects of the polymer matrix are enhanced due to the incorporation of nanoclay [114, 115]. These nanoclay based materials have been used as cost effective substituents with significant strength characteristics and considerable enhancement of properties on the nanoclay reinforced polymer-based hybrid composite systems. Several studies have been carried out on nanoclay reinforced polymer composites [113–119]. For instance; Hussain et al. [120] have reported that the natural fiber composite is robustly dependent on the optimum fiber length and weight percentage

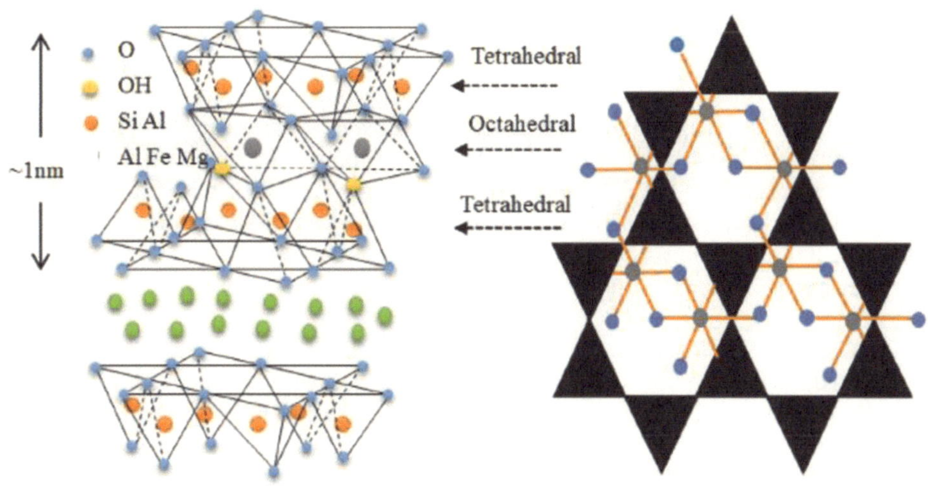

Figure 2.
Schematic illustration of the montmorillonite clay (MMT) structure. Side view of tetrahedron units of MMT assembled through weak van der Waals and electrostatic forces to form the primary particles and top view of MMT shows hexagonal structure of oxygen and hydroxyl ligands of the octahedral layer. Reprinted with permission from Ref. [111].

of fiber. It is also reported that the mechanical properties of the polymer composite system are improved by the incorporation of a little amount of nanoclay in to the matrix [121].

3. Synthesis of carbon-based nano-filler reinforced epoxy composites

Several methods have been employed for the synthesis of various nano-filler reinforced epoxy resin composites in recent few times. In this section, we have emphasized mainly on three methods (including in-situ intercalative polymerization, solution mixing and melt blending) for preparation of composites, which are extensively discussed in details.

3.1 *In-situ* intercalative polymerization method

In situ polymerization is a very effective method that allows carbon-based filler particles to be uniformly dispersed in the matrix and therefore gives strong interaction between polymeric matrix and reinforcing filler in the composite systems. This technique generally involves the polymerization of monomers in the presence of dispersed filler particles. The polymerization is initiated through the incorporation of filler and suitable initiator with the application of heat or radiation after diffusion during the synthesis of the polymeric materials [122, 123]. Several researchers have been reported about the preparation of composites by using *in situ* polymerization methods and achieving superior mechanical properties with the low value of percolation threshold as compared to the other techniques (like solution compounding or melt blending method) [124–127]. However, in-situ polymerization is also been used to give non-covalent composites based on different polymeric matrices including polyethylene (PE) [128] and PMMA [129], etc. Ray and Okamoto [130] reported that the *in-situ* polymerization method referred as intercalation polymerization method is applied for the preparation of GDs based nanocomposites. Zhang et al. [124] have fabricated graphene sheet reinforced epoxy resin composites by in-situ polymerization method. It is observed that the prepared composites have significant improvement in the Young's modulus and thermal stability at 0.7 wt% of GNs content in the matrix. It is also noticed that there is an extensive enhancement in the thermal conductivity of the resultant graphene sheet-epoxy composite systems using in-situ polymerization method [131]. Ying and his co-workers [132, 133] have fabricated surfactant incorporated CNT into the epoxy composites *via* in-situ polymerization technique. It is observed that the tensile strength and ultimate strain are enhanced with an increase in the modified filler contents in the CNT reinforced epoxy composite systems.

Kotsilkova et al. [134] have reported on amine and epoxy grafted MWCNTs based composites. They have successfully synthesized and measured various performances (including rheological, dc conductivity, radio frequency and microwave properties) of the composites. The enhancement of the thermal and mechanical properties of the MWCNTs reinforced epoxy composites were utilized by in-situ polymerization method, which is reported by Theodore et al. [135]. The preparation of the epoxy-based nanocomposites using functionalized vapor grown carbon nanofiber (VGCNF) by in-situ polymerization technique. It is revealed that the functionalized VGCNF-epoxy composite systems have superior tensile modulus and strength than that of the neat VGCNF and epoxy matrix with enhanced thermal stability [136]. The carbon nanofiber (CNF) based epoxy composites with various weight percentages of silane-modified CNF and unmodified CNF contents *via* in-situ polymerization technique were prepared. It is revealed that the synthesized composites with different properties (thermal, mechanical and electrical) were analyzed as a function of different weight percentages of modified and unmodified CNF contents [66, 137]. The key features of this technique offer covalent bonding between surface-functionalized CNTs and polymeric matrix, which result enhancement of mechanical performance of the composites with strong interfacial bond [86, 138]. On the other hand, the shortcomings of this method are that the dispersion of filler particles in the polymeric matrix occurred during the preparation of nanocomposites, so the plenty of energy is necessary, which might affect the large production of polymer-based nanocomposites. Similarly, Ramezanzadeh et al. [139] have prepared conducting polypyrrole (PPy) and zinc doped polypyrrole functionalized graphene oxide (GO) nano-sheets for the development of high-performance epoxy-based composite systems with enhanced thermal and mechanical performance. The results showed that the nanocomposites are enhanced about 54% for elongation at break and 115% for the energy at break, and there is an improvement of tensile strength and energy (21% and 32.44%) at break for the zinc doped GO-PPy in the epoxy composites.

3.2 Solution mixing/solution blending/solvent casting

Solution-induced intercalation technique is a simple, ease of processing and efficient approach to produce polymer-based nanocomposites. This technique is more suitable for both small- and large-scale preparation of the nanocomposites. In this technique, initially an appropriate proportion of polymeric mixtures were dissolved in a proper solvent. Then the different types of filler particles (such as graphene, carbon nanotubes (CNTs), carbon nanofibers (CNFs), etc.) at various weight percentages are reinforced into the polymeric solution with constant stirring either by magnetic stirrer or mechanically or by using ultrasonication technique for dissolution of the particles in the matrix. Once the complete mixing of solution is over, then the mixtures were dried to eliminate the solvent and ultimately resultant composite materials are molded with suitable mold to give necessary shape and making suitable for characterization and further properties measurement. Similarly, the solvent casting method has been used to synthesize composites in the form of thin films. This method deals with the uniform dissolution of ingredients in a suitable solvent and then evaporated through a drying device. However, the solvent casting technique shows higher mixing quality, thinner film, high purity and better clarity than that of the melt mixing technique [140]. Besides, the solution or polymer film is exposed to moderately low thermal or mechanical stress during the preparation technique, which results insignificant degradation or side reactions.

There are various literatures available with epoxy-based composites by using this technique and then studied several properties of those composites as a function of different weight percentage of filler contents. Similar to epoxy resin, other various polymers (such as PS, PMMA, PVDF and PI) and its co-polymers [P(VDF-HFP), P(VDF-TrFE), etc.] are used to fabricate nanocomposites with the addition of several fillers into matrix using this technique [141–146]. For instance; Lv et al. [147], Prolongo et al. [148] and Allaoui et al. [64] have fabricated carbon nano-fibers reinforced epoxy-based composites with different weight percentages of CNF contents and studied their properties. Similarly, Choi et al. [149] have synthesized polycarbonate-based carbon nanofiber composites. They have also examined mechanical properties and electrical resistivity with the incorporation of various percentages of CNTs contents. Moreover, the limitation of this technique is the requirement of the high amount of solvent in the preparation of nanocomposites for industrial point of view, which may not be environmentally friendly and cost-effective.

3.3 Melt blending method

Melt blending is the facile, cost-effective, eco-friendly and conventional technique for the synthesis of various thermoplastic polymer-based composites. In this process, there is no such solvent used for preparation. Also, it is one of the most efficient approaches for industries in large scale production of the composite materials. During the last few times, most of the industrial researchers have preferred this synthesis technique for preparing nanocomposites. In this technique, initially, polymeric materials and filler particles are normally mixed systematically and then the homogeneous mixture is subjected to annealing greater than the melting point of the polymeric material. Besides, the polymeric material is melted and combined with apposite amount of the filler particles using an extruder. The process of melt blending is carried out in the presence of an inert gas including argon, nitrogen or neon. However, this melt blending technique has enormous benefits over in-situ intercalative polymerization or polymer solution intercalation method. This technique is compatible with present industrial processes including extrusion and injection molding techniques and thus it is a most popular method. Various academic and industrial researchers have fabricated different polymer-based nanocomposites (PP-CNT, nylon 6-MWCNT, PC-MWCNT and PS-MWCNTs, etc.) using this technique [150–153]. For instance; Jin and his co-workers [154] have fabricated PMMA-MWCNTs based composites by using this technique. In the synthesized composite system CNTs shows no obvious damage or breakage as well as uniformly dispersed within the polymeric matrix. The preparation of polylactide (PLA) exfoliated graphite (EG) based nanocomposites was synthesized using this technique. It is seen that the composite exhibits a considerable increase in the thermal degradation temperature with the increase in the EG contents. The mechanical and electrical performances are also improved with the continuous increase of graphite contents. A similar work on carbon-based material (such as graphite nano-sheet and MWCNTs, etc.) reinforced epoxy composites is fabricated by Kim et al. and his group [155, 156] using this technique and also studied their mechanical and rheological properties of the resultant composites.

4. Different properties of filler reinforced epoxy composites

Recently, epoxy-based composites are of great importance for their significant dielectric, electrical, thermal and mechanical performances. In this section, we have especially emphasized on properties (dielectric, electrical and mechanical) of the various filler reinforced epoxy-based composites.

4.1 Dielectric and electrical properties of epoxy-based polymer composites

Varma et al. [157] have fabricated calcium copper titanate ($CaCu_3Ti_4O_{12}$; CCTO) and metallic aluminum (Al) powder-filled epoxy-based tri-phase composites with various weight percentages of filler contents. It is observed that the dielectric constant of the composites is greatly improved near the percolation threshold. The maximum dielectric constant was achieved (≈700) of the three-phase epoxy composites, which is much larger than that of the two-phase epoxy-CCTO (≈70) and neat epoxy matrix. These flexible three-phase composites are potential candidates for practical application in the field of energy storage devices [158]. The perovskite-type ceramic ($BaTiO_3$) based epoxy composites with a different weight percentage of filler contents have been reported by Kuo et al. [159]. They studied the dielectric properties of these composites and obtained a high dielectric constant (≈44) and negligible dielectric loss (<1) at 40 wt% of filler content in the epoxy matrix. The synthesized three-phase composites have been uniformly dispersed in the polymeric matrix. However, the dielectric constant of these composites is proportional to the volume ratio of the ceramic contents and remained constant with the application of temperature and frequencies. The fabricated composites have larger than that of the commercial ceramic-filled polymer composites. Bhattacharya and Tummala [160] have fabricated PMN-PT filled epoxy composites with various weight percentages of filler contents. It is revealed that the high dielectric constant (≈29) was achieved in the 40 wt% of filler-filled epoxy composites with superior homogeneity within the epoxy matrix. The dielectric performance of Ni particle doped epoxy composites at 40 and 55 wt% of filler contents have been analyzed in the various frequency regions from 1 to 10^7 Hz and temperature range of −20 to 200°C. The dielectric constant of these Ni particle filled epoxy composites is improved with the increase of Ni particle content and simultaneously reduces the frequency at room temperature. The value of the dielectric constant is increased with the increase of filler contents due to the improved dipole and interfacial polarization effect. On the other hand, the dielectric loss value was reduced with the higher concentration of filler contents has reported by Chen and his co-workers [161]. The multifunctional polymer-based composites comprising of Fe_3O_4 fillers and epoxy resin as the matrix was analyzed in different concentration of magnetite Fe_3O_4 contents. It is observed that the composites with larger dielectric constant and suppressed dielectric loss at low frequencies. However, the synthesized composites with a larger volume fraction of filler contents appeared at the percolation threshold with an increase in the value of conductivity [162]. Xie et al. [163] have fabricated the modified hexagonal boron nitride reinforced epoxy composites with enhanced dielectric properties *via* simple free radical polymerization technique. It is observed that PGMA grafted h-BN reinforced epoxy matrix composite has improved thermal conductivity with increase of filler contents. The modified composite shows larger dielectric constant than that of the unmodified one and pristine epoxy matrix. Similarly, the dielectric loss of the resultant composites was achieved negligible at suitable frequencies. Zhang et al. [164] have fabricated core-shell satellite structured $BaTiO_3$ nanoparticles with polydopamine (PDA) layers and silver (Ag) nanoparticles incorporated into the epoxy matrix. It is observed that the epoxy nanocomposites with BT-PDA and BT-PDA-Ag fillers showed enhanced dielectric constant (≈9) and negligible dielectric loss (0.024) at microwave frequencies (10 GHz) for 20 vol% of filler contents. However, it is also noticed that the composite shows homogeneous dispersion with uniform particle size, due to strong interfacial interaction between modified particles and polymer matrix. Meng [165] and his co-workers have developed thermally stable honokiol derived epoxy resin nanocomposites with excellent thermal and dielectric properties. The results showed that the fabricated composites have excellent dielectric constant (9.74) and minimized dielectric loss values (0.026) at 1 KHz. Moreover, it is also confirmed the better thermal stability, thermal conductivity and high specific heat in the composites.

4.2 Mechanical properties of epoxy-based composites

Zhao et al. [166] have synthesized hyper branched graphene oxide structured based epoxy nanocomposites. The results showed excellent engineering application performances of the composite systems. The synthesized nanocomposites have uniform dispersion in the epoxy resin matrix and are combined with the matrix through chemical bonds, which shows strong interfacial active force and enhancing the load transfer efficiency of the matrix to hyper-branched polymer-graphene composites. However, these composites exhibited excellent mechanical properties [impact strength (58.53%), tensile strength (83.29%), and compression strength (57%)] with considerable increase of 0.2 wt% for HPB-GO contents than that of the neat epoxy matrix. Also, it is noticed that there is 80% increase in thermal conductivity (0.32 W m^{-1} K^{-1}) of the synthesized nanocomposites. The incorporation of rGO in the epoxy matrix in the composites improved the strength and Young's modulus is about 500 and 70%. It is observed that rGO plays a significant role in strengthening the epoxy than that of the GO. However, rGO efficiency in the improvement of modulus and strength is about 10–35% than those of GO. The significant results for r-GO composite are due to the efficiency of interfacial adhesion between r-GO sheet and epoxy molecules [167]. The enhancement of strength can be ascribed to the outstanding load bearing capacity of reduced GO sheets as well as excellent load transfer from matrix to reinforcements. They also suggest an increase in the value of modulus and strength in the composites for 1 wt% rGO reinforced polymeric matrix. The fabrication of graphene- nano-alumina based epoxy composites reported by Osman and his co-workers [168]. It is observed that both tensile strength and storage modulus are improved by 22.56% and 4.6%, which is much larger in contrast to the neat epoxy matrix. The thermal conductivity of the resultant composites is also improved by 23.4% with increase in the filler contents. The incorporation of alumina particles on the surface of the graphene not only reduces the electron transfer but also eliminates the agglomeration of graphene. Khan [169] and his co-workers have reported improvement of thermo-mechanical performance of carbon fiber and glass fiber reinforced epoxy composites. These composites were characterized by using universal testing machine (UTM) with tensile strength and Young's modulus. The tensile strength of carbon fiber reinforced epoxy composites is enhanced to 844.44%, 951.11% and 1122.22% with selected 40, 50 and 60 wt% of carbon fiber contents. On the other hand, the tensile strength of glass fiber reinforced epoxy composites is also enhanced about 156.66%, 171.10% and 197.77% for 40, 50 and 60 wt% of glass fiber contents. Karle et al. [170] have systematically studied $CaSiO_3$ particulate fillers reinforced epoxy-based composites and analyzed their hardness, flexural strength and impact resistance through mechanical performance. It is found that the addition of $CaSiO_3$ particles (1–2 wt%) in the matrix results effective enhancement of mechanical properties than that of the neat epoxy. Park [171] and his co-workers have investigated the reinforcement effect of molybdenum sulfide (MoS_2) nano-sheets on the mechanical performance of the epoxy-based composites. The fabricated high performance epoxy composites are extremely enhanced the fracture toughness (55–81%), flexural strength (25–66%), modulus (0.7–6%), impact strength (31–118%) and strong interfacial interaction (1–21% surface free energy) than that of the pristine epoxy matrix. The thermal and mechanical performance of the epoxy reinforced modified iron oxide nanoparticles reported by Baghdadi et al. [172]. It is revealed that the PDA modified Fe_3O_4 based epoxy composites normally improved as compared to neat epoxy matrix and unmodified one. The maximum enhancement in tensile strength (34%) and fracture toughness (13%) is observed in the epoxy-based composites. The graphene oxide-epoxy composites with improved failure strength (48.3%) and toughness (1185.2%) for 0.0375 wt% of GO within the epoxy matrix. The fabricated composites with enhanced properties may be due to the uniform dispersion of the GO in the epoxy matrix through two-phase extraction technique using an aqueous dispersion of the GO were reported by Yang and his co-workers [173]. Bortz et al. [174] have reported helical carbon nanofibers to achieve graphene nanoribbon and then oxidized to get GO. This composite exhibit improved tensile strength (7.57 MPa) and modulus (3.32 GPa) at 0.5 and 0.1 wt% of GO loading into the polymer matrix. Moreover, the flexural strength was also improved by 12% and 23% for the addition of fillers into the matrix.

Munoz et al. [175] have synthesized GO based epoxy composites and studied their mechanical properties for various wt% of GO contents. It is revealed that the composites with enhanced elastic modulus and flexural modulus at 0.3 wt% of GO contents in the epoxy matrix. Fang et al. [176] have reported methylene dianiline (MDA) modified rGO-epoxy composites with different weight percentages of filler contents. In these composite systems, the fracture toughness and flexural strength is enhanced by 94% and 92% for 0.6 wt% filler loading. Seong et al. [177] have fabricated MDA modified GNP-epoxy composites and investigated their mechanical properties. The 1.5 phr MDA modified GNP content shows 120% and 63% enhancement in impact toughness and storage modulus in the epoxy-based composites. Naebe [178] and his co-workers have reported the effect of thermally reduced graphene oxide (TRG) with bingel reaction (FG) based epoxy composites. It is observed that 0.1 wt% of filler content in the FG-epoxy composite shows larger flexural strength (15% and 22%) and higher storage modulus (6% and 16%) than that of the pristine epoxy matrix. Similarly, Guo et al. [179] have synthesized GO modified triazine derivatives (GO-TCT-DETA) show homogeneous dispersion in the epoxy matrix. It is also exhibited higher flexural strength (49%) and modulus (15%) for 0.1 wt% of GO-TCT-DETA-based epoxy composites than that of the pristine epoxy matrix.

5. Summary

The epoxy-based composites have become potential candidates for the application in different technological fields owing to their excellent physical, chemical and electrical performances. The present chapter reviews on the research work related to carbon based nanofillers (e.g., fullerene, CNTs, CNFs, graphene & nanoclay) reinforced epoxy-based composites. However, several researchers have reported the variety of reinforcement in order to explore mechanical, dielectric & electrical performances to predict their behavior suitable for technological viability of the epoxy-based composites. According to the literatures, it has been found that the use of nanofillers into the epoxy matrix, which help to enhance the mechanical and electrical performance of the composite systems than that of the micron sized filler particles. Moreover, with the increase of concentration of nano-filler content results in the enhancement of certain properties of the epoxy-based composites up to the threshold value, in most of the cases after attaining the threshold value the properties again initiate to fall, this may be ascribed to the presence of large clusters into the epoxy resin. The incorporation of minimum amount of nano-filler contents is indispensible into the matrix to avoid agglomeration in the composite systems. Therefore, we have especially emphasized on most suitable synthesis techniques for the processing of nano-filler reinforced epoxy-based composites. Further, our efforts have been made for providing higher insight on the properties (like dielectric, electrical and mechanical) of various nanofillers reinforced epoxy composites. The unique properties of nanofiller reinforced epoxy composites also reflected as potential applicability in the field of electronics, energy storage, gas sensors and aerospace.

Acknowledgements

The authors gratefully acknowledge the support provided by Centurion University of Technology and Management, Odisha, India for carrying out the present research work.

Conflicts of interest

The authors declare no conflict of interest.

Author details

Srikanta Moharana[1*] and Bibhuti B. Sahu[2*]

1 School of Applied Sciences, Centurion University of Technology and Management, Odisha, India

2 Department of Physics, Veer Surendra Sai University of Technology, Odisha, India

*Address all correspondence to: srikantanit@gmail.com; bibhubhusan78@gmail.com

References

[1] Luz GM, Mano JF. Mineralized structures in nature: Examples and inspirations for the design of new composite materials and biomaterials. Composites Science and Technology. 2010;**70**:1777-1788

[2] Song K, Zhang Y, Meng J, Green EC, Tajaddod N, Li H, et al. Structural polymer-based carbon nanotube composite fibers: Understanding the processing–structure–performance relationship. Maternité. 2013;**6**:2543-2577

[3] Bello SA, Agunsoye JO, Hassan SB, Zebase Kana MG, Raheem IA. Epoxy resin based composites, mechanical and Tribological properties: A review. Tribology in Industry. 2015;**37**:500-524

[4] Alarifi IM. Investigation the conductivity of carbon fiber composites focusing on measurement techniques under dynamic and static loads. Journal of Materials Research and Technology. 2019;**8**:4863-4893

[5] Chandgude S, Salunkhe S. In state of art: Mechanical behavior of natural fiber-based hybrid polymeric composites for application of automobile components. Polymer Composites. 2021;**42**:2678-2703

[6] de Leon AC, Chen Q, Palaganas NB, Palaganas JO, Manapat J, Advincula RC. High performance polymer nanocomposites for additive manufacturing applications. Reactive and Functional Polymers. 2016;**103**:141-155

[7] Njuguna J, Pielichowski K, Alcock JR. Epoxy-based fibre reinforced nanocomposites. Advanced Engineering Materials. 2007;**9**:835-847

[8] Day RJ, Lovell PA, Wazzan AA. Toughened carbon/epoxy composites made by using core/shell particles. Composites Science and Technology. 2001;**61**:41-56

[9] Jiao J, Sun X, Pinnavaia TJ. Mesostructured silica for the reinforcement and toughening of rubbery and glassy epoxy polymers. Polymer. 2009;**50**:983-989

[10] Gu H, Ma C, Gu J, Guo J, Yan X, Huang J, et al. An overview of multifunctional epoxy nanocomposites. Journal of Materials Chemistry C. 2016;**4**:5890-5906

[11] Hadavand BS, Javid KM, Gharagozlou M. Mechanical properties of multi-walled carbon nanotube/epoxy polysulfide nanocomposite. Materials and Design. 2013;**50**:62-67

[12] PalDey SCDS, Deevi SC. Single layer and multilayer wear resistant coatings of (Ti, Al) N: A review. Materials Science and Engineering A. 2003;**342**:58-79

[13] Samal P, Vundavilli PR, Meher A, Mahapatra MM. Recent progress in aluminum metal matrix composites: A review on processing, mechanical and wear properties. Journal of Manufacturing Processes. 2020;**59**:131-152

[14] Surappa MK. Aluminium matrix composites: Challenges and opportunities. Sadhana. 2003;**28**:319-334

[15] Perez Ipina JE, Yawny AA, Stuke R, Gonzalez Oliver C. Fracture toughness in metal matrix composites. Materials Research. 2000;**3**:74-78

[16] Lynch S. A review of underlying reasons for intergranular cracking for a variety of failure modes and materials and examples of case

histories. Engineering Failure Analysis. 2019;**100**:329-350

[17] Rawal SP. Metal-matrix composites for space applications. JOM. 2001;**53**: 14-17

[18] Kurumlu D, Payton EJ, Young ML, Schöbel M, Requena G, Eggeler G. High-temperature strength and damage evolution in short fiber reinforced aluminum alloys studied by miniature creep testing and synchrotron microtomography. Acta Materialia. 2012;**60**:67-78

[19] Abbah J, Amos S, Chindo B, Ngazal I, Vongtau HO, Adzu B, et al. Pharmacological evidence favouring the use of Nauclea latifolia in malaria ethnopharmacy: Effects against nociception, inflammation, and pyrexia in rats and mice. Journal of Ethnopharmacology. 2010;**127**:85-90

[20] Teli MD, Pandit P. Development of thermally stable and hygienic colored cotton fabric made by treatment with natural coconut shell extract. Journal of Industrial Textiles. 2018;**48**:87-118

[21] Guimarães JL, Frollini E, Da Silva CG, Wypych F, Satyanarayana KG. Characterization of banana, sugarcane bagasse and sponge gourd fibers of Brazil. Industrial Crops and Products. 2009;**30**:407-415

[22] Sekhon SS, Kaur P, Park JS. From coconut shell biomass to oxygen reduction reaction catalyst: Tuning porosity and nitrogen doping. Renewable and Sustainable Energy Reviews. 2021;**2021**(147):111173

[23] Dwivedi C, Manjare S, Rajan SK. Recycling of waste tire by pyrolysis to recover carbon black: Alternative & environment-friendly reinforcing filler for natural rubber compounds. Composites Part B: Engineering. 2020;**200**:108346

[24] Martínez JD, Cardona-Uribe N, Murillo R, García T, López JM. Carbon black recovery from waste tire pyrolysis by demineralization: Production and application in rubber compounding. Waste Management. 2019;**85**:574-584

[25] Bashir AS, Manusamy Y. Recent developments in biocomposites reinforced with natural biofillers from food waste. Polymer—Plastics Technology and Engineering. 2015;**54**:87-99

[26] Jose AS, Athijayamani A, Jani SP. A review on the mechanical properties of bio waste particulate reinforced polymer composites. Materials Today: Proceedings. 2021;**37**:1757-1760

[27] Bledzki AK, Mamun AA, Volk J. Barley husk and coconut shell reinforced polypropylene composites: The effect of fibre physical, chemical and surface properties. Composites Science and Technology. 2010;**70**:840-846

[28] Atuanya CU, Aigbodion VS, Nwigbo SC. Experimental study of the thermal and wear properties of recycled polyethylene/breadfruit seed hull ash particulate composites. Materials and Design. 2014;**53**:65-73

[29] Zhang X, Zhao N, He C. The superior mechanical and physical properties of nanocarbon reinforced bulk composites achieved by architecture design—A review. Progress in Materials Science. 2020;**113**:100672

[30] Naseer A, Ahmad F, Aslam M, Guan BH, Harun WSW, Muhamad N, et al. A review of processing techniques for graphene-reinforced metal matrix composites. Materials and Manufacturing Processes. 2019;**34**: 957-985

[31] Wang G, Yu D, Kelkar AD, Zhang L. Electrospun nanofiber: Emerging reinforcing filler in polymer matrix composite materials. Progress in Polymer Science. 2017;**75**:73-107

[32] Zhang Y, Li Y, Ma H, Yu T. Tensile and interfacial properties of unidirectional flax/glass fiber reinforced hybrid composites. Composites Science and Technology. 2013;**88**:172-177

[33] Wetzel B, Haupert F, Zhang MQ. Epoxy nanocomposites with high mechanical and tribological performance. Composites Science and Technology. 2003;**63**:2055-2067

[34] Rehman MM, Zeeshan M, Shaker K, Nawab Y. Effect of micro-crystalline cellulose particles on mechanical properties of alkaline treated jute fabric reinforced green epoxy composite. Cellulose. 2019;**26**:9057-9069

[35] Saba N, Jawaid M, Alothman OY, Paridah MT, Hassan A. Recent advances in epoxy resin, natural fiber-reinforced epoxy composites and their applications. Journal of Reinforced Plastics and Composites. 2016;**35**:447-470

[36] Saba N, Jawaid M, Paridah MT, Al-Othman OY. A review on flammability of epoxy polymer, cellulosic and non-cellulosic fiber reinforced epoxy composites. Polymers for Advanced Technologies. 2016;**27**:577-590

[37] Kumar K, Ghosh PK, Kumar A. Improving mechanical and thermal properties of TiO_2-epoxy nanocomposite. Composites Part B: Engineering. 2016;**97**:353-360

[38] Yousri OM, Abdellatif MH, Bassioni G. Effect of Al_2O_3 nanoparticles on the mechanical and physical properties of epoxy composite. Arabian Journal for Science and Engineering. 2018;**43**: 1511-1517

[39] Inam F, Heaton A, Brown P, Peijs T, Reece MJ. Effects of dispersion surfactants on the properties of ceramic–carbon nanotube (CNT) nanocomposites. Ceramics International. 2014;**40**:511-516

[40] Namitha LK, Chameswary J, Ananthakumar S, Sebastian MT. Effect of micro-and nano-fillers on the properties of silicone rubber-alumina flexible microwave substrate. Ceramics International. 2013;**39**:7077-7087

[41] Abenojar J, Martínez MA, Velasco F, Pascual-Sánchez V, Martín-Martínez JM. Effect of boron carbide filler on the curing and mechanical properties of an epoxy resin. The Journal of Adhesion. 2009;**85**:216-238

[42] Shi G, Zhang MQ, Rong MZ, Wetzel B, Friedrich K. Sliding wear behavior of epoxy containing nano-Al_2O_3 particles with different pretreatments. Wear. 2004;**256**:1072-1081

[43] Bhagyashekar MS, Rao RMVGK. Effects of material and test parameters on the wear behavior of particulate filled composites part 1: SiC-epoxy and gr-epoxy composites. Journal of Reinforced Plastics and Composites. 2007;**26**:1753-1768

[44] Bazrgari D, Moztarzadeh F, Sabbagh-Alvani AA, Rasoulianboroujeni M, Tahriri M, Tayebi L. Mechanical properties and tribological performance of epoxy/Al_2O_3 nanocomposite. Ceramics International. 2018;**44**:1220-1224

[45] Thakur AS, Sharma N, Kango S, Sharma S. Effect of nanoparticles on epoxy based composites: A short review. Materials Today: Proceedings. 2021;**44**:4640-4642

[46] Rodgers RM, Mahfuz H, Rangari VK, Chisholm N, Jeelani S. Infusion of SiC nanoparticles into SC-15 epoxy: An investigation of thermal and mechanical response. Macromolecular Materials and Engineering. 2005;**290**:423-429

[47] Paul R, Dai L. Interfacial aspects of carbon composites. Composite Interfaces. 2018;**25**:539-605

[48] Kroto HW, Heath JR, O'Brien SC, Curl RF, Smalley RE. C60: Buckminsterfullerene. Nature. 1985;**318**:162-163

[49] Sasidharan S, Anand A. Epoxy-based hybrid structural composites with nanofillers: A review. Industrial and Engineering Chemistry Research. 2020;**59**:12617-12631

[50] Iijima S. Helical microtubules of graphitic carbon. Nature. 1991;**354**:56-58

[51] Iijima S, Ichihashi T. Single-shell carbon nanotubes of 1-nm diameter. Nature. 1993;**363**:603-605

[52] Novoselov KS, Geim AK, Morozov SV, Jiang DE, Zhang Y, Dubonos SV, et al. Electric field effect in atomically thin carbon films. Science. 2004;**306**:666-669

[53] Bu G, Ou Z, Li Z, Liu F, Zheng H, Zou X, et al. Static and buckling characteristics of the porous ring reinforced by graphene nanofillers. Engineering Structures. 2022;**251**:113536

[54] Zhang W, Chai C, Fan Q, Song Y, Yang Y. Two-dimensional carbon allotropes with tunable direct band gaps and high carrier mobility. Applied Surface Science. 2021;**537**:147885

[55] Zhu Y, Murali S, Cai W, Li X, Suk JW, Potts JR, et al. Graphene and graphene oxide: Synthesis, properties, and applications. Advanced Materials. 2010;**22**:3906-3924

[56] Joshi G, Mehta KP. Fabrication and applications of fullerene-based metal nanocomposites: A review. Journal of Materials Research. 2021;**36**:114-128

[57] Zhang M, Li J. Carbon nanotube in different shapes. Materials Today. 2009;**12**:12-18

[58] Ruoff RS, Qian D, Liu WK. Mechanical properties of carbon nanotubes: Theoretical predictions and experimental measurements. Comptes Rendus Physique. 2003;**4**:993-1008

[59] Kathi J, Rhee KY. Surface modification of multi-walled carbon nanotubes using 3-aminopropyltriethoxysilane. Journal of Materials Science. 2008;**43**:33-37

[60] Chen H, Jacobs O, Wu W, Rüdiger G, Schädel B. Effect of dispersion method on tribological properties of carbon nanotube reinforced epoxy resin composites. Polymer Testing. 2007;**26**:351-360

[61] Karousis N, Tagmatarchis N, Tasis D. Current progress on the chemical modification of carbon nanotubes. Chemical Reviews. 2010;**110**:5366-5397

[62] Kim JA, Seong DG, Kang TJ, Youn JR. Effects of surface modification on rheological and mechanical properties of CNT/epoxy composites. Carbon. 2006;**44**:1898-1905

[63] Montazeri A, Chitsazzadeh M. Effect of sonication parameters on the mechanical properties of multi-walled carbon nanotube/epoxy composites. Materials and Design. 2014;**56**:500-508

[64] Allaoui A, Bai S, Cheng HM, Bai JB. Mechanical and electrical properties of a MWNT/epoxy composite.

Composites Science and Technology. 2002;**62**:1993-1998

[65] Montazeri A, Khavandi A, Javadpour J, Tcharkhtchi A. Viscoelastic properties of multi-walled carbon nanotube/epoxy composites using two different curing cycles. Materials and Design. 2010;**31**:3383-3388

[66] Shen J, Huang W, Wu L, Hu Y, Ye M. Thermo-physical properties of epoxy nanocomposites reinforced with amino-functionalized multi-walled carbon nanotubes. Composites Part A: Applied Science and Manufacturing. 2007;**38**:1331-1336

[67] Choi JR, Rhee KY, Park SJ. Influence of electrolessly silver-plated multi-walled carbon nanotubes on thermal conductivity of epoxy matrix nanocomposites. Composites Part B: Engineering. 2015;**80**:379-384

[68] Innocenzi P, Brusatin G. Fullerene-based organic–inorganic nanocomposites and their applications. Chemistry of Materials. 2001;**13**:3126-3139

[69] Rafiee MA, Yavari F. Fullerene-Epoxy nanocomposites-enhanced mechanical properties at low nanofiller loading. Journal of Nanoparticle Research. 2011;**13**:733-737

[70] Kausar A. Advances in polymer/fullerene nanocomposite: A review on essential features and applications. Polymer-Plastics Technology and Materials. 2017;**56**:594-605

[71] Kausar A. Fullerene nanofiller reinforced epoxy nanocomposites—Developments, progress and challenges. Materials Research Innovations. 2021;**25**:175-185

[72] Ogasawara T, Ishida Y, Kasai T. Mechanical properties of carbon fiber/fullerene-dispersed epoxy composites. Composites Science and Technology. 2009;**69**:2002-2007

[73] Jiang Z, Zhang H, Zhang Z, Murayama H, Okamoto K. Improved bonding between PAN-based carbon fibers and fullerene-modified epoxy matrix. Composites. Part A, Applied Science and Manufacturing. 2008;**39**:1762-1767

[74] Manawi YM, Samara A, Al-Ansari T, Atieh MA. A review of carbon nanomaterials' synthesis via the chemical vapor deposition (CVD) method. Maternité. 2018;**11**:822

[75] Ribeiro H, Schnitzler MC, da Silva WM, Santos AP. Purification of carbon nanotubes produced by the electric arc-discharge method. Surfaces and Interfaces. 2021;**26**:101389

[76] Gupta N, Gupta SM, Sharma SK. Carbon nanotubes: Synthesis, properties and engineering applications. Carbon Letters. 2019;**29**:419-447

[77] Wu HY, Jia LC, Yan DX, Gao JF, Zhang XP, Ren PG, et al. Simultaneously improved electromagnetic interference shielding and mechanical performance of segregated carbon nanotube/polypropylene composite via solid phase molding. Composites Science and Technology. 2018;**156**:87-94

[78] Razavi R, Zare Y, Rhee KY. The roles of interphase and filler dimensions in the properties of Tunneling spaces between CNT in polymer nanocomposites. Polymer Composites. 2019;**40**:801-810

[79] Soni SK, Thomas B, Kar VR. A comprehensive review on CNTs and CNT-reinforced composites: Syntheses, characteristics and applications. Materials Today Communications. 2020;**25**:101546

[80] Mohd Nurazzi N, Asyraf MM, Khalina A, Abdullah N, Sabaruddin FA, Kamarudin SH. & Sapuan SM. Fabrication, functionalization, and application of carbon nanotube-reinforced polymer composite: An overview. Polymer. 2021;**13**:1047

[81] Zhang D, Ye K, Yao Y, Liang F, Qu T, Ma W, et al. Controllable synthesis of carbon nanomaterials by direct current arc discharge from the inner wall of the chamber. Carbon. 2019;**142**:278-284

[82] Subramani M, Ramamoorthy M. Vibration analysis of multiwalled carbon nanotube-reinforced composite shell: An experimental study. Polymers and Polymer Composites. 2020;**28**:223-232

[83] Azizi-Lalabadi M, Hashemi H, Feng J, Jafari SM. Carbon nanomaterials against pathogens; the antimicrobial activity of carbon nanotubes, graphene/graphene oxide, fullerenes, and their nanocomposites. Advances in Colloid and Interface Science. 2020;**284**:102250

[84] Li Y, Huang X, Zeng L, Li R, Tian H, Fu X, et al. A review of the electrical and mechanical properties of carbon nanofiller-reinforced polymer composites. Journal of Materials Science. 2019;**54**:1036-1076

[85] Liu S, Chevali VS, Xu Z, Hui D, Wang H. A review of extending performance of epoxy resins using carbon nanomaterials. Composites. Part B, Engineering. 2018;**136**:197-214

[86] Haghgoo M, Ansari R, Hassanzadeh-Aghdam MK. Predicting effective electrical resistivity and conductivity of carbon nanotube/carbon black-filled polymer matrix hybrid nanocomposites. Journal of Physics and Chemistry of Solids. 2022;**161**:110444

[87] Gkikas G, Barkoula N, Paipetis AS. Effect of dispersion conditions on the Thermo-mechanical and toughness properties of multi walled carbon nanotubes-reinforced epoxy. Composites. Part B, Engineering. 2012;**43**:2697-2705

[88] Sham ML, Kim JK. Surface functionalities of multi-wall carbon nanotubes after UV/ozone and TETA treatments. Carbon. 2006;**44**:768-777

[89] Xin F, Li L. Decoration of carbon nanotubes with silver nanoparticles for advanced CNT/polymer nanocomposites. Composites. Part A, Applied Science and Manufacturing. 2011;**42**:961-967

[90] Guadagno L, Raimondo M, Vittoria V, Vertuccio L, Lafdi K, De Vivo B, et al. The role of carbon nanofiber defects on the electrical and mechanical properties of CNF-based resins. Nanotechnology. 2013;**24**:305704

[91] Deeraj BDS, George G, Dhineshbabu NR, Bose S, Joseph K. Electrospun ZrO_2@ carbon nanofiber mats and their epoxy composites as effective EMI shields in Ku band. Materials Research Bulletin. 2021;**144**:111477

[92] Park SY, Yook S, Goo S, Im W, Youn HJ. Preparation of transparent and thick CNF/epoxy composites by controlling the properties of cellulose nanofibrils. Nanomaterials. 2020;**10**:625

[93] Lavagna L, Nistico R, Musso S, Pavese M. Functionalization as a way enhance dispersion of carbon nanotubes in matrices: A review. Materials Today Chemistry. 2021;**20**:100477

[94] Dong M, Zhang H, Tzounis L, Santagiuliana G, Bilotti E, Papageorgiou DG. Multifunctional epoxy nanocomposites reinforced by two-dimensional materials: A review. Carbon. 2021;**185**:57-81

[95] Zhang Z, Du J, Li J, Huang X, Kang T, Zhang C, et al. Polymer nanocomposites with aligned two-dimensional materials. Progress in Polymer Science. 2021;**114**:101360

[96] Sahu D, Sutar H, Senapati P, Murmu R, Roy D. Graphene, graphene-derivatives and composites: Fundamentals, synthesis approaches to applications. Journal of Composites Science. 2021;**5**:181

[97] Sreedhar I, Upadhyay U, Roy P, Thodur SM, Patel CM. Carbon capture and utilization by graphenes-path covered and ahead. Journal of Cleaner Production. 2021;**284**:124712

[98] Park M, Kim N, Lee J, Gu M, Kim BS. Versatile graphene oxide nanosheets via covalent functionalization and their applications. Materials Chemistry Frontiers. 2021;**5**:4424-4444

[99] Lee SY, Mahajan RL. A facile method for coal to graphene oxide and its application to a biosensor. Carbon. 2021;**181**:408-420

[100] Rathinavel S, Priyadharshini K, Panda D. A review on carbon nanotube: An overview of synthesis, properties, functionalization, characterization, and the application. Materials Science and Engineering B. 2021;**268**:115095

[101] Kooti M, Sedeh AN, Gheisari K, Figuerola A. Synthesis, characterization, and performance of nanocomposites containing reduced graphene oxide, polyaniline, and cobalt ferrite. Physica B: Condensed Matter. 2021;**612**:412974

[102] Alexopoulos ND, Paragkamian Z, Poulin P, Kourkoulis SK. Fracture related mechanical properties of low and high graphene reinforcement of epoxy nanocomposites. Composites Science and Technology. 2017;**150**:194-204

[103] Rafiee MA, Rafiee J, Wang Z, Song H, Yu Z, Koratkar N. Enhanced mechanical properties of nanocomposites at low graphene content. ACS Nano. 2009;**3**:3884-3890

[104] Jang BZ, Zhamu A. Processing of nanographene platelets (NGPs) and NGP nanocomposites: A review. Journal of Materials Science. 2008;**43**:5092-5101

[105] Wazalwar R, Sahu M, Raichur AM. Mechanical properties of aerospace epoxy composites reinforced with 2D nano-fillers: Current status and road to industrialization. Nanoscale Advances. 2021;**3**:2741-2776

[106] Razaq A, Bibi F, Zheng X, Papadakis JSHM, Li H. Review on graphene-, graphene oxide-, reduced graphene oxide-based flexible composites: From fabrication to applications. Maternité. 2022;**15**:1012

[107] Gantayat S, Prusty G, Rout DR, Swain SK. Expanded graphite as a filler for epoxy matrix composites to improve their thermal, mechanical and electrical properties. New Carbon Materials. 2015;**30**:432-437

[108] Dehghani A, Bahlakeh G, Ramezanzadeh B. Synthesis of a non-hazardous/smart anti-corrosion nano-carrier based on beta-cyclodextrin-zinc acetylacetonate inclusion complex decorated graphene oxide (β-CD-ZnA-MGO). Journal of Hazardous Materials. 2020;**398**:122962

[109] Ahmad K. Study of different polymer nanocomposites and their pollutant removal efficiency. Polymer. 2021;**217**:123453

[110] Abulyazied DE, Ene A. An investigative study on the progress of nanoclay-reinforced polymers: Preparation, properties, and

applications: A review. Polymer. 2021;**13**: 4401

[111] Zhu TT, Zhou CH, Kabwe FB, Wu QQ, Li CS, Zhang JR. Exfoliation of montmorillonite and related properties of clay/polymer nanocomposites. Applied Clay Science. 2019;**169**:48-66

[112] Krasinskyi V, Suberlyak O, Sikora J, Zemke V. Nanocomposites based on polyamide-6 and montmorillonite intercalated with polyvinylpyrrolidone. Polymer-Plastics Technology and Materials. 2021;**2021**(60):1641-1655

[113] Karki S, Gohain MB, Yadav D, Ingole PG. Nanocomposite and bio-nanocomposite polymeric materials/ membranes development in energy and medical sector: A review. International Journal of Biological Macromolecules. 2021;**193**:2121-2139

[114] Rajeshkumar G, Seshadri SA, Ramakrishnan S, Sanjay MR, Siengchin S, Nagaraja KC. A comprehensive review on natural fiber/ nano-clay reinforced hybrid polymeric composites: Materials and technologies. Polymer Composites. 2021;**42**:3687-3701

[115] Jawaid M, Chee SS, Asim M, Saba N, Kalia S. Sustainable kenaf/bamboo fibers/clay hybrid nanocomposites: Properties, environmental aspects and applications. Journal of Cleaner Production. 2022;**330**:129938

[116] Pumchusak J, Thajina N, Keawsujai W, Chaiwan P. Effect of organo-modified montmorillonite nanoclay on mechanical, thermo-mechanical, and thermal properties of carbon fiber-reinforced phenolic composites. Polymer. 2021;**13**:754

[117] Tay CH, Mazlan N, Sultan MTH, Abdan K, Lee CH. Mechanical performance of hybrid glass/kenaf epoxy composite filled with organomodified nanoclay. Journal of Materials Research and Technology. 2021;**15**:4415-4426

[118] Shettar M, Kowshik CS, Manjunath M, Hiremath P. Experimental investigation on mechanical and wear properties of nanoclay–epoxy composites. Journal of Materials Research and Technology. 2020;**9**:9108-9116

[119] Zawawi EZE, Hafizah AN, Romli AZ, Yuliana NY, Bonnia NN. Effect of nanoclay on mechanical and morphological properties of poly (lactide) acid (PLA) and polypropylene (PP) blends. Materials Today: Proceedings. 2021;**46**:1778-1782

[120] Barbhuiya AH, Ismail K. Effect of fiber length and loading on the properties of Schumannianthus dichotomus (murta) fiber–reinforced epoxy composites. International Journal of Polymer Analysis and Characterization. 2016;**21**:221-227

[121] Mylsamy B, Palaniappan SK, Subramani SP, Pal SK, Aruchamy K. Impact of nanoclay on mechanical and structural properties of treated Coccinia indica fibre reinforced epoxy composites. Journal of Materials Research and Technology. 2019;**8**:6021-6028

[122] Zheng W, Lu X, Wong SC. Electrical and mechanical properties of expanded graphite-reinforced high-density polyethylene. Journal of Applied Polymer Science. 2004;**91**:2781-2788

[123] Liang J, Wang Y, Huang Y, Ma Y, Liu Z, Cai J, et al. Electromagnetic interference shielding of graphene/epoxy composites. Carbon. 2009;**47**:922-925

[124] Zhang Z, Zhang W, Li D, Sun Y, Wang Z, Hou C, et al. Mechanical and anticorrosive properties of graphene/ epoxy resin composites coating prepared

by in-situ method. International Journal of Molecular Sciences. 2015;**16**:2239-2251

[125] Khan T, Irfan MS, Ali M, Dong Y, Ramakrishna S, Umer R. Insights to low electrical percolation thresholds of carbon-based polypropylene nanocomposites. Carbon. 2021;**176**:602-631

[126] Naeem M, Kuan HC, Michelmore A, Yu S, Mouritz AP, Chelliah SS, et al. Epoxy/graphene nanocomposites prepared by in-situ microwaving. Carbon. 2021;**177**:271-281

[127] Luo X, Schubert DW. Examining the contribution of factors affecting the electrical behavior of poly (methyl methacrylate)/graphene nanoplatelets composites. Journal of Applied Polymer Science. 2021;**138**:50694

[128] Fim FDC, Guterres JM, Basso NR, Galland GB. Polyethylene/graphite nanocomposites obtained by in situ polymerization. Journal of Polymer Science, Part A: Polymer Chemistry. 2010;**48**:692-698

[129] Jang JY, Kim MS, Jeong HM, Shin CM. Graphite oxide/poly(methyl methacrylate) nanocomposites prepared by a novel method utilizing macroazoinitiator. Composites Science and Technology. 2009;**69**:186-191

[130] Ray SS, Okamoto M. Polymer/layered silicate nanocomposites: A review from preparation to processing. Progress in Polymer Science. 2003;**28**:1539-1641

[131] Das TK, Prusty S. Graphene-based polymer composites and their applications. Polymer-Plastics Technology and Engineering. 2013;**52**:319-331

[132] Du JH, Ying Z, Bai S, Li F, Sun C, Cheng HM. Microstructure and resistivity of carbon nanotube and nanofiber/epoxy matrix nanocomposites. International Journal of Nanoscience. 2002;**1**:719-723

[133] Ying Z, Du JH, Bai S, Li F, Liu C, Cheng HM. Mechanical properties of surfactant-coating carbon nanofiber/epoxy composite. International Journal of Nanoscience. 2002;**1**:425-430

[134] Kotsilkova R, Ivanov E, Bychanok D, Paddubskaya A, Demidenko M, Macutkevic J, et al. Effects of sonochemical modification of carbon nanotubes on electrical and electromagnetic shielding properties of epoxy composites. Composites Science and Technology. 2015;**106**:85-92

[135] Theodore M, Hosur M, Thomas J, Jeelani S. Influence of functionalization on properties of MWCNT–epoxy nanocomposites. Materials Science and Engineering. 2011;**528**:1192-1200

[136] Wang DH, Sihn S, Roy AK, Baek JB, Tan LS. Nanocomposites based on vapor-grown carbon nanofibers and an epoxy: Functionalization, preparation and characterization. European Polymer Journal. 2010;**46**:1404-1416

[137] Zhu J, Wei S, Ryu J, Budhathoki M, Liang G, Guo Z. In situ stabilized carbon nanofiber (CNF) reinforced epoxy nanocomposites. Journal of Materials Chemistry. 2010;**20**:4937-4948

[138] Ma PC, Siddiqui NA, Marom G, Kim JK. Dispersion and functionalization of carbon nanotubes for polymer-based nanocomposites: A review. Composites Part A: Applied Science and Manufacturing. 2010;**41**:1345-1367

[139] Saadatmandi S, Ramezanzadeh B, Asghari M, Bahlakeh G. Graphene oxide nanoplatform surface decoration by spherical zinc-polypyrrole nanoparticles for epoxy coating properties

enhancement: Detailed explorations from integrated experimental and electronic-scale quantum mechanics approaches. Journal of Alloys and Compounds. 2020;**816**:152510

[140] Tang ZG, Black RA, Curran JM, Hunt JA, Rhodes NP, Williams DF. Surface properties and biocompatibility of solvent-cast poly [ε-caprolactone] films. Biomaterials. 2004;**25**:4741-4748

[141] Li Z, Zhang Y, Liang S, Li G. Effect of surfactant on microstructure, surface hydrophilicity, mechanical and thermal properties of different multi-walled carbon nanotube/polystyrene composites. Materials Research Express. 2018;**5**:055035

[142] Roshan MJ, Jeevika A, Bhattacharyya A, Shankaran DR. One-pot fabrication and characterization of graphene/PMMA composite flexible films. Materials Research Bulletin. 2018;**105**:133-141

[143] Moharana S, Mahaling RN. Enhancement investigations on dielectric and electrical properties of niobium pentoxide (Nb_2O_5) reinforced poly (vinylidene fluoride)(PVDF)-graphene oxide (GO) nanocomposite films. Journal of Asian Ceramic Societies. 2021;**9**:1183-1193

[144] Moharana S, Mahaling RN. Silver (Ag)-graphene oxide (GO)-poly (vinylidene fluoride-co-hexafluoropropylene)(PVDF-HFP) nanostructured composites with high dielectric constant and low dielectric loss. Chemical Physics Letters. 2017;**680**:31-36

[145] Dai T, Yu Z, Yuan S, Zou H, Liu P. Gradient structure polyimide/graphene composite aerogels fabricated by layer-by-layer assembly and unidirectional freezing. Journal of Applied Polymer Science. 2021;**138**:50153

[146] Wu Q, Guo H, Sun H, Liu X, Sui H, Wang F. Flexible piezoelectric energy harvesters with graphene oxide nanosheets and PZT-incorporated P(VDF-TrFE) matrix for mechanical energy harvesting. Ceramics International. 2021;**47**:19614-19621

[147] Xiao LV, Yang S, Jin J, Li LZG, Jiang J. Preparation and electromagnetic properties of carbon nanofiber/epoxy composites. Journal of Macromolecular Science, Part B. Physics. 2010;**49**:355-365

[148] Prolongo SG, Campo M, Gude MR, Moran RC, Urena A. Thermophysical characterization of epoxy resin reinforced by amino-functionalized carbon nanofibers. Composites Science and Technology. 2009;**69**:349-357

[149] Choi YK, Sugimoto KI, Song SM, Endo M. Production and characterization of polycarbonate composite sheets reinforced with vapor grown carbon fiber. Composites. 2006;**37**:1944-1951

[150] Ma Y, Wu D, Liu Y, Li X, Qiao H, Yu ZZ. Electrically conductive and super-tough polypropylene/carbon nanotube nanocomposites prepared by melt compounding. Composites. 2014;**56**:384-391

[151] Liu TX, Phang IY, Shen L, Chow SY, Zhang WD. Morphology and mechanical properties of multiwalled carbon nanotubes reinforced Nylon-6 composites. Macromolecules. 2004;**37**:7214-7222

[152] Poetschke P, Bhattacharya AR, Janke A, Goering H. Melt mixing of polycarbonate/multi-wall carbon nanotube composites. Composite Interfaces. 2003;**10**:389-404

[153] Li Y, Pionteck J, Pötschke P, Voit B. Thermal annealing to influence the vapor sensing behavior of co-continuous poly

(lactic acid)/polystyrene/multiwalled carbon nanotube composites. Materials and Design. 2020;**187**:108383

[154] Jin ZX, Pramoda KP, Xu GQ, Goh SH. Dynamic mechanical behavior of melt-processed multi-walled carbon nanotube/poly(methyl methacrylate) composites. Chemical Physics Letters. 2001;**337**:43-47

[155] Kim IH, Jeong YG. Polylactide/exfoliated graphite nanocomposites with enhanced thermal stability, mechanical modulus, and electrical conductivity. Journal of Polymer Science. 2010;**48**:850-858

[156] Kim KS, Choi KE, Park SJ. Electrical and mechanical properties of graphite Nanosheet/carbon nanotubes-filled epoxy nanocomposites. Carbon Letters. 2009;**10**:335-338

[157] Prakash BS, Varma KBR. Dielectric behavior of CCTO/epoxy and Al-CCTO/epoxy composites. Composites Science and Technology. 2007;**67**:2363-2368

[158] Bai Y, Cheng ZY, Bharti V, Xu HS, Zhang QM. High-dielectric constant ceramic-powder polymer composites. Applied Physics Letters. 2000;**25**:3804-3806

[159] Kuo DH, Chang CC, Su TY, Wang WK, Lin BY. Dielectric properties of three ceramic/epoxy composites. Materials Chemistry and Physics. 2004;**85**:201-206

[160] Bhattacharya SK, Tummala RR. Integral passives for next generation of electronic packaging: Application of epoxy/ceramic nanocomposites as integral capacitors. Microelectronics Journal. 2001;**32**:11-19

[161] Wang Z, Zhou W, Dong L, Sui X, Cai H, Zuo J, et al. Dielectric spectroscopy characterization of relaxation process in Ni/epoxy composites. Journal of Alloys and Compounds. 2016;**682**:738-745

[162] Ramajo LA, Cristóbal AA, Botta PM, López JP, Reboredo MM, Castro MS. Dielectric and magnetic response of Fe_3O_4/epoxy composites. Composites Part A: Applied Science and Manufacturing. 2009;**40**(4):388-393

[163] Jiang Y, Shi X, Feng Y, Li S, Zhou X, Xie X. Enhanced thermal conductivity and ideal dielectric properties of epoxy composites containing polymer modified hexagonal boron nitride. Composites Part A: Applied Science and Manufacturing. 2018;**107**:657-664

[164] Zhang M, Xu P, Peng H, Qin F. A rational design of core-shell-satellite structured $BaTiO_3$ fillers for epoxy-based composites with enhanced microwave dielectric constant and low loss. Composites Part B: Engineering. 2021;**215**:108764

[165] Meng J, Chen P, Yang R, Dai L, Yao C, Fang Z, et al. Thermal stable honokiol-derived epoxy resin with reinforced thermal conductivity, dielectric properties and flame resistance. Chemical Engineering Journal. 2021;**412**:128647

[166] Zhao Y, Wu Z, Guo S, Zhou Z, Miao Z, Xie S, et al. Hyperbranched graphene oxide structure-based epoxy nanocomposite with simultaneous enhanced mechanical properties, thermal conductivity, and superior electrical insulation. Composites Science and Technology. 2022;**217**:109082

[167] Kausar A, Rafique I, Anwar Z, Muhammad B. Perspectives of epoxy/graphene oxide composite: Significant features and technical applications. Polymer—Plastics Technology and Engineering. 2016;**55**:704-722

[168] Osman A, Elhakeem A, Kaytbay S, Ahmed A. Thermal, electrical and mechanical properties of graphene/nano-alumina/epoxy composites. Materials Chemistry and Physics. 2021;**257**:123809

[169] Khan ZI, Arsad A, Mohamad Z, Habib U, Zaini MAA. Comparative study on the enhancement of thermo-mechanical properties of carbon fiber and glass fiber reinforced epoxy composites. Materials Today: Proceedings. 2021;**39**:956-958

[170] Karle AH, Nukulwar MR, Tungikar VB. Evaluation of mechanical and thermal properties of epoxy composites reinforced with $CaSiO_3$ particulate fillers. Materials Today: Proceedings. 2021;**46**:325-330

[171] Riaz S, Park SJ. A comparative study on nanoinclusion effect of MoS_2 nanosheets and MoS2 quantum dots on fracture toughness and interfacial properties of epoxy composites. Composites Part A: Applied Science and Manufacturing. 2021;**146**:106419

[172] Baghdadi YN, Youssef L, Bouhadir K, Harb M, Mustapha S, Patra D, et al. Thermal and mechanical properties of epoxy resin reinforced with modified iron oxide nanoparticles. Journal of Applied Polymer Science. 2021;**138**:50533

[173] Yang H, Shan C, Li F, Zhang Q, Han D, Niu L. Convenient preparation of tunably loaded chemically converted graphene oxide/epoxy resin nanocomposites from graphene oxide sheets through two-phase extraction. Journal of Materials Chemistry. 2009;**19**:8856-8860

[174] Bortz DR, Heras EG, Martin-Gullon I. Impressive fatigue life and fracture toughness improvements in graphene oxide/epoxy composites. Macromolecules. 2012;**45**:238-245

[175] Horta Muñoz S, Serna Moreno MDC, González-Domínguez JM, Morales-Rodríguez PA, Vázquez E. Experimental, numerical, and analytical study on the effect of graphene oxide in the mechanical properties of a solvent-free reinforced epoxy resin. Polymer. 2019;**11**:2115

[176] Fang M, Zhang Z, Li J, Zhang H, Lu H, Yang Y. Constructing hierarchically structured interphases for strong and tough epoxy nanocomposites by amine-rich graphene surfaces. Journal of Materials Chemistry. 2010;**20**:9635-9643

[177] Seong M, Kim DS. Effects of facile amine-functionalization on the physical properties of epoxy/graphene nanoplatelets nanocomposites. Journal of Applied Polymer Science. 2015;**132**:42269

[178] Naebe M, Wang J, Amini A, Khayyam H, Hameed N, Li LH, et al. Mechanical property and structure of covalent functionalised graphene/epoxy nanocomposites. Scientific Reports. 2014;**4**:1-7

[179] Guo S, Ma L, Song G, Li X, Li P, Wang M, et al. Covalent grafting of triazine derivatives onto graphene oxide for preparation of epoxy composites with improved interfacial and mechanical properties. Journal of Materials Science. 2018;**53**:16318-16330

Section 3

Applications

Chapter 5

Epoxy Resin Adhesives: Modification and Applications

Jun Zhang, Hai Luo, Xiaojian Zhou and Bowen Liu

Abstract

Epoxy resin adhesives (ERAs) as easily prepared thermosetting adhesives have been extensively employed in building construction, electrical appliance manufacturing, automobile manufacturing and wood industry because of their excellent mechanical properties, water resistance, low cost, long service life and strong bonding properties. This chapter aims to introduce the synthesis, properties and development of ERAs and to illustrate how defects in their curing properties, thermal properties, brittleness and flammability affect their global development. Furthermore, this study introduces the modification of ERAs according to these defects and their development in main application fields. Lastly, the limitations and prospects of ERAs in future applications are also discussed.

Keywords: epoxy resin, adhesives, mechanical properties, curing properties, modification

1. Introduction

Epoxy resin adhesive (ERA) is the general term for polymers with two or more epoxy groups in the molecule. It is widely used in industries and is an important thermosetting resin adhesive [1, 2]. ERA is a thermosetting adhesive with strong adhesion, high cohesion, low shrinkage, low cost and low creep rate. It can be used for several materials, such as metal, cement and wood; thus, it is referred to as a 'universal and strong' glue [3]. It has a history of more than 70 years. The molecular end of epoxy resin is connected with epoxy groups. During curing, hydroxyl and ether bonds are formed, and the structure contains benzene or heterocyclic rings. Due to the presence of epoxy groups, hydroxyl groups, ether bonds, ester groups and other polar groups, it has a strong bonding effect on many substances other than non-polar polymers [4]. After the epoxy-based adhesive is cured, it forms a complex three-dimensional (3D) network structure with strong cohesion. The epoxy-based adhesive hardly generates low-molecular products during curing, has a small linear expansion coefficient, stable dimensions, small internal stress and better bonding strength. Epoxy-based adhesives meet the requirements of structural adhesives, but they also have some shortcomings. Because the curing process of epoxy-based adhesives needs a higher temperature, and it contains many rigid groups, such as a benzene or heterocyclic ring, the flexibility of the molecular chain is minimal [5].

Besides, after cross-linking to form a network structure, the deformability is further weakened, showing strong brittleness, which results in low bonding strength, poor impact strength, delamination and easy cracking resulting of epoxy-based adhesives [3]. Meanwhile, its flame retardancy is poor. As a structural adhesive, it is expected to cure quickly, have higher heat resistance and flame retardancy. Therefore, epoxy-based adhesives must be modified to expand its scope of application. This study mainly introduces the curing, heat resistance, toughening and flame-retardant modification of epoxy-based adhesives and their application in different fields.

2. Modification of ERAs

2.1 Curing modification of epoxy adhesives by curing agent

As a thermosetting adhesive, ERA must be cured at high temperatures. However, for industrial applications, it must be cured at room temperature [6, 7], so the importance of developing curing agents for epoxy-based adhesive cured at room temperature is self-evident. Especially, room-temperature fast-curing epoxy adhesives can be used in aerospace and marine engineering applications, as well as in traditional manufacturing and daily life [8, 9], because of their fast-curing speed, high strength and strong durability [10]. With the continuous development of curing agents, room-temperature fast-curing ERAs as chemical products have become indispensable in the manufacturing industry [11]. The room-temperature curing of ERAs is an energy-saving curing method. The curing process is simple, and it is suitable for various curing situations that do not require heating.

2.1.1 Classification of curing agent for ERAs

According to the curing temperature, epoxy-based adhesive curing agents can be divided into amines, acid anhydrides, synthetic resins and latent curing agents by different chemical components. Among them, amine-curing agents are often used in ambient curing at room temperature [12]. Amine-curing agents are the earliest room-temperature curing agent used. It adheres excellently to most adherents. However, amine-curing agents have high volatility and toxicity and have strong water and carbon dioxide absorption abilities. The cured surface is prone to whitening and blistering [13]. Among the amine-curing agents, there are mainly polyamides, aliphatic amines, alicyclic amines. In industrial applications, curing agents, such as aliphatic amines, polyamides and alicyclic amines, are often used [14].

2.1.1.1 Polyamide-curing agent

As one commonly used curing agent, polyamide accounts for more than 30% of the total epoxy resin-curing agents. It is mainly made of dimers or unsaturated fatty acids and polyamine as raw materials and forms amide bonds through dehydration condensation [15]. Among them, the dimerised fatty acid polyamide can overcome the shortcomings of epoxy-based adhesives' fragility and has low toxicity, good workability and high-paint film adhesion [16]. Modifying the polyamide-curing agent can effectively improve the properties of epoxy resins. For example, Bryan et al. [17] used polyamide and phthalic anhydride as the curing agent of epoxy resin to improve the curing rate at room temperature. Gholipour et al. [18] improved the thermal properties of epoxy resins by preparing polyamidoamine (PAMAM) dendrimer-curing agents grafted with graphene oxide.

2.1.1.2 Aliphatic amine-curing agent

The amount of aliphatic amine-curing agents in various curing agents is second only to polyamide because most are liquid and have good miscibility with epoxy resin. Epoxy resin can be cured at room temperature. Modifying the aliphatic amine-curing agent can effectively improve the mechanical properties of epoxy resin. For example, Patel et al. [19] brominated unsaturated castor oil, which was the main raw material, and reacted the resulting material with excess aliphatic diamines, such as ethylenediamine, 1,3-propanediamine and 1,6-hexanediamine, to obtain an amino-functionalised castor-oil-curing agent to improve the mechanical strength of epoxy resins. Wan et al. [20] synthesised a novel low-volatility star aliphatic polyamine with extremely high-NH_2 functional groups as the curing agent of bisphenol A diglycidyl ether epoxy resin. The novel curing agent has a high reaction activity, and the reaction has autocatalytic properties. Additionally, compared with linear propylene diamine, it can significantly increase the crosslinking density and glass transition temperature (Tg) of the cured epoxy resin.

2.1.1.3 Alicyclic amine-curing agent

Alicyclic amines are amine compounds containing alicyclic rings. The alicyclic amine-curing agent has many spatial conformations and good flexibility. Most alicyclic amines are low-viscosity liquids with long pot life and excellent chroma and gloss. Alicyclic polyamine compounds are widely used as curing agents for epoxy-resin adhesives and other structure adhesives because the molecular structure contains alicyclic rings (five-membered or six-membered rings) with higher stiffness and better stability. Xu et al. [21] used alicyclic polyamines and acrylonitrile to synthesise the curing agent to improve the bonding strength of epoxy-resin-based adhesives.

2.2 Heat-resistant modification of ERAs

ERAs can withstand high temperatures up to 175°C and are compatible with all common reinforcement materials. A higher-temperature-resistant ERA can be essentially applied in many fields. In addition to having high-temperature-resistant properties, it can also show strong properties in many aspects, such as high-temperature-resistant epoxy. It has excellent mechanical properties, relatively outstanding strength and has good corrosion resistance and insulation properties [22, 23]. Therefore, research on high-temperature-resistant ERA is extremely necessary. There are two main measures in implementing the modification of high-temperature ERA. The first measure entails introducing new structures into the epoxy resin itself to improve its high-temperature properties. The second measure is blending or co-polymerisation to modify the high-temperature epoxy resin.

2.2.1 ERA itself introduces a new structure to improve its high-temperature resistance properties

The modification treatment of high-temperature-resistant ERA is mainly performed to promote the structure of ERA to be changed to a certain extent. It is more common to improve the high-temperature resistance effect by introducing new structures. This method for introducing a new structure through the epoxy resin also involves many types of processing in actual implementation. For example, the effective use of multifunctional structures can promote the formation of ring structures [24]. The functionality of the epoxy-resin structure increases, which can promote the stability and cross-linking density of the corresponding structure and finally effectively enhance the high-temperature resistance of the epoxy resin. Furthermore, introducing rigid groups that have a good high-temperature resistance effect can promote the epoxy resin to show excellent performance, such as benzene ring, fused ring and biphenyl are some of the more commonly used rigid groups [25]. Moreover, Bismaleimide and epoxy resin may form an interpenetrating network or two-phase system during the polymerisation process, which improves the toughness and heat resistance of the epoxy resin [26].

Luo et al. [27] modified bisphenol A epoxy resin with bismaleimide and 4,4′-diaminodiphenylsulfone to produce a two-component high-strength bismaleimide modified epoxy-based adhesive with high cross-linking. The viscosity of the adhesive gradually decreases as temperature increases and can maintain good mechanical properties and storage stability. Cheng et al. [28] used 2,7-dihydroxynaphthalene and epi-chlorohydrin as raw materials to synthesise an epoxy-resin-based adhesive containing a naphthalene ring structure. Also, Yang et al. [29] used 1-naphthol and dicyclopentadiene as the main raw materials to synthesise an ERA containing naphthalene ring and dicyclopentadiene structure. The results show that the ERA has a higher heat resistance than the bisphenol A epoxy resin.

2.2.2 Modification of high-temperature-resistant ERA by blending or co-polymerisation

The blending and co-polymerisation methods can effectively and mainly help select an ideal material and ERA for effective combination to ensure greater high-temperature-resistant properties. Combined with the specific application and implementation of these methods, the requirements for specific blended or co-polymerised materials are relatively strict [6]. For example, the appropriate use of heat-resistant polymers, nano-materials and silicones can achieve ideal modifications. The treatment effect improves the heat resistance of the epoxy resin; besides, it can also guarantee its toughness or strength to be ideally optimised. Zhang et al. [30] modified ordinary bisphenol A epoxy resin with organic silicon active intermediates, added nitrile-40 and nano-TiO_2 active fillers to toughen and strengthen the resin. The results showed that the modified ERA that could be cured at room temperature, used for a long time at 250°C and can withstand 300°C for a short period has been developed. Hu et al. [31] used polymethyltriethoxysilane (PTS) to react with a synthetic phosphorus-containing silane coupling agent in a certain ratio to modify the bisphenol A epoxy resin. The modified ERA retained its tensile strength. However, the Tg, high-temperature thermal stability, impact strength and limiting oxygen index (LOI) were all improved. Ramirez et al. [32] combined epoxycyclohexyldimethylsilyl with the curing agent 4,4′-(1,3-phenylene diisopropylidene) diphenylamine after coordination. Due to the dispersion of the heat-resistant epoxycyclohexyldimethylsilyl in the ERA, the silicon oxide compound formed during the thermal decomposition process deposited on the surface of the unburned polymer, partially forming a protective layer, slowing down the heat transfer to a certain extent and inhibiting the flammability as the gas volatilises, thereby preventing the mixture of flammable gas and oxygen.

2.3 Toughening modification of ERAs

ERAs are cost-effective and have simple moulding and processing methods, low chemical shrinkage after curing, good chemical stability, excellent mechanical properties and good bonding properties [33]. However, due to several epoxy groups, the cured structure has a high chemical cross-link density, low-molecular chain flexibility and high internal stress, resulting in greater brittleness, poor impact resistance and fatigue resistance of the ERA. It limits its application and development in some high-tech fields that require high durability and reliability.

ERA has good compatibility with rubber and other elastomers. After the rubber is dissolved in the uncured epoxy-resin matrix, the ERA undergoes a curing reaction, separates from the rubber and is dispersed in the resin to form a 'sea island' structure, thereby improving the toughness of the epoxy resin [34]. The rubber molecules containing no reactive groups cannot react with epoxy resin and will precipitate out during curing, which has a toughening effect. However, if excessively added, it weakens the adhesion of the bonding interface. Therefore, rubber molecules with active groups are generally used to modify and toughen epoxy resins. Carboxyl-terminated liquid nitrile rubber (CTBN) and amino-terminated liquid nitrile rubber (ATBN) have been widely mixed with epoxy resins to improve their toughness. For example, Wang et al. [35] used CTBN and ERA to prepare a structural adhesive with high shear and excellent peel strength. Meanwhile, to adapt to the application in different fields and improve the toughness of the cured ERA, flexible segments are often introduced into the curing agent to control its physical and chemical properties. Lou et al. [36] used dendritic polyester polyol as the branching unit and toughening segment and imidazole-terminated diisocyanate as a functional group to synthesise a functional toughening-curing agent to improve the toughness of ERAs. In the initial curing stage, several secondary hydroxyl groups react with the isocyanate groups to form a dendritic epoxy structure, and the bisphenol A epoxy molecule acts as a long-chain polyol to react with the dendritic epoxy structure. Simultaneously, the epoxy group opens a ring to form a secondary hydroxyl group and continuously reacts with the isocyanate group in the cross-linking structure. Furthermore, NH- existing in the carbamate reacts with the epoxy group, thereby obtaining an epoxy resin-curing cross-linking system with a 3D dendritic cross-linking structure. Thus, the toughness of epoxy resin has been greatly improved. Meanwhile, Zheng et al. [37] studied the toughening effect of nano-SiO_2 on cycloaliphatic epoxy systems. They used nanosilica to improve the toughness of cycloaliphatic ERA. The coupling agent γ-glycid oxypropyltrimethoxysilane (KH-560) was used to modify the surface properties of SiO_2. The results show that adding nano-SiO_2 effectively improves the toughness and thermal stability of the cycloaliphatic ERA.

2.4 Flame-retardant modification of ERA

ERAs have been widely used due to their excellent properties [38]. However, the conventional ERA is formed from reacting bisphenol A and epi-chlorohydrin [39]. It is flammable when cured, which could cause a high fire risk when ERAs are used in certain applications, such as printed manufacture, furniture, aircraft and train interiors [40]. Serious consequences could occur due to the high release rate of heat and smoke accompanied by the combustion of epoxy resins. Therefore, it is very important to enhance the flame retardancy of ERA for expanding their application in this field. Many studies have reported improving the flame retardancy of epoxy resins via structural modification or adding various flame retardants [41, 42]. Structural modification introduces the elements with flame-retardant functions into the molecular structure of ERAs. The representative of structural modification is brominated ERA, which is the reaction product of epi-chlorohydrin and brominated bisphenol A, such as tetrabromo diphenylolpropane. The brominated ERA has an outstanding flame ignition resistance, whereas the bromine content is ~18–20% in the finished adhesive. When the product is thermally decomposed at the temperature generated in the fire, it will release acid halide gas, which protects the product from fire. These halide gases act as extinguishers to significantly increase the ignition temperature of the cured ERA. The addition of flame retardants in ERA shows good properties, such as simple processing, low cost, wide source of raw materials and obvious flame-retardant effect. It is one of the most popular strategies for flame-retardant modification of ERA. Flame retardants could be an integral part of the ERA by reacting chemically with the polymers or simply mixed with the ERA without any reaction. All kinds of flame retardants work by acting chemically and/or physically either in the vapour phase and/or condensed phase to interfere with the combustion process during heating, pyrolysis, ignition or flame spread [43]. The types of flame retardants and their operating characteristics are described as follows [38]: (1) char formers: usually, phosphorus compounds, which remove the carbon fuel source and provide an insulation layer against the fire's heat. (2) Heat absorbers: usually metal hydrates, such as aluminium trihydrate (ATH) or magnesium hydroxide, which remove heat by evaporating the water in their structure. (3) Flame quenchers: usually, bromine- or chlorine-based halogen systems that interfere with the reactions in a flame. (4) synergists: Usually, antimony compounds, which enhance the performance of the flame quencher. The flame-retarding action of ERA could be divided into physical and chemical actions. Physical action includes cooling, barrier action via the formed protective layer and fuel dilution. For the cooling action, flame retardants absorb the heat when they decompose, and the endothermic decomposition may consume the released heat from the combustion of ERA, then the burning adhesive is cooled. Generally, most inorganic-hydrated compounds, such as aluminium and magnesium hydroxides, may play a role via this mode. For the barrier action, the decomposition products of some flame retardants shield the surface of the adhesive and form a protective layer that may act as a barrier to resist oxygen and the produced heat. Consequently, the burning process is difficult to sustain. For fuel dilution, some flame retardants may release water vapour, carbon dioxide, or other inert gases, thereby decreasing the concentration of free radicals and combustible gases in the burning adhesive. Chemical action includes gas-phase and condensed-phase reactions. The gas-phase reaction mechanism is generally regarded as the interruption of the chain reaction of the ERA structural system during burning. The flame retardant that provides the flame-retarding action via the gas-phase reaction action may capture free radicals to decrease the concentration of free radicals than the combustion threshold and then prevent or delay burning, in which halogen-containing flame retardants are the most representative. During burning, the halogen-containing flame retardants release the hydrogen halide, which may react with the free radicals formed during burning to inhibit the combustion of substrates. For the condensed-phase reaction, the flame retardant that provides the flame-retarding action via the condensed-phase reaction may promote the formation of a carbonised or vitreous layer by cross-linking, aromatising, catalytic dehydration of polymers or reacting with the ERA. In this flame-retardant mode, intumescent flame retardants may form an intumescent char layer by some chemical reactions during burning, and generally, the formed char layer may promote the barrier action and improve the flame retardancy of the ERA. Furthermore, some flame retardants can accelerate the rupture of the chains of the ERA, and several droplets are produced under this condition. Then, a large amount of heat may be taken away when these droplets move away from the burning zone [44]. Flame retardants can be classified into several families, including halogen-based compounds, phosphorus-based compounds, silicon-based compounds, nano-composites and metal-based compounds. Among them, halogen-based and phosphorus-based flame retardants are widely used.

For halogen-containing flame retardants, the flammability of ERA can be greatly reduced by incorporating a halogen into the molecule. The best known are the halogen-containing ERAs based on chlorinated, brominated and fluorinated bisphenol A. They often comprise blends of two or more epoxy-resin systems, one of which is a halogenated resin, and the other of which perhaps contains a halogenated curing agent, such as chloric anhydride. These halogen-containing ERAs have been developed over decades and are still used widely due to the obvious advantages of low cost, processability, miscibility and low reduction in physical/mechanical features of the flame-retardant systems. Halogen-containing flame retardants function by liberating acid halide gases as the product thermally breaks down at the high temperatures incurred in a fire. These halide gases act as extinguishers to significantly increase the ignition temperature of the cured ERA. The mechanism of these flame retardants is the release of hydrogen halides (HCl and HBr) during the thermal decomposition of the ERA. The chemical reaction during burning is a free radical chain reaction, and the continuous growth of free radicals is important for maintaining the burning process for the ERA. Several chemical halide intermediates form during the burning of ERAs. These halide species are carried into the flame front of the burning polymer where they inhibit key free radical reactions of combustion. This inhibition results in flames becoming unsteady and extinguishing and lowers the release of heat overall [10, 11]. Generally, alicyclic or aliphatic halogen-containing flame retardants are more efficient than aromatic halogen compounds. Alicyclic or aliphatic halogen-containing flame retardants burn at low temperatures for most polyolefins because of lower carbon-halogen bond energies and easier halogen release [45]. Beach et al. [46] synthesised brominated polybutadiene-polystyrene (BrPBPS) flame retardant from styrene-butadiene-styrene triblock architecture by bromination. The BrPBPS flame retardant contains similar aliphatic bromine as in hexabromocyclododecane, but with a higher-molecular-weight structure. It provides similar flame-retardant activity as hexabromocyclododecane in polystyrene blends, where both release HBr to provide the gas-phase activity. Both also provide enhanced ERA degradation as another major pathway for condensed flame-retardant activity. Jiang et al. [47] added BrPBPS into epoxy asphalt adhesive to enhance its flame resistance. Meanwhile, the Tg of the epoxy asphalt adhesive was notably enhanced with the inclusion of BrPBPS. Wu et al. [48] synthesised liquid-oxygen-compatible bromine-containing ERA by the polycondensation of tetrabromobisphenol A and epoxy resins. The bromine element was introduced into the ERA to improve the liquid oxygen compatibility and enhance flame retardancy. The results showed that limiting oxygen index increased drastically when the bromine content was increased from 0% to 21.20%.

Phosphorus-containing flame retardants are identified as one of the most promising halogen-free flame retardants [49, 50] since they possess excellent properties, such as low-smoke emission, low toxicity, form a stable carbonised layer after burning effectively [44, 46, 51, 52] and are environmentally friendly [35]. For preparing organophosphorus epoxides-based adhesive, three general methods were employed.

First, the condensation of 1-chloro-2,3-epoxypropane and organophosphorus compounds containing two or more hydroxyl groups. Second, the Michaelis-Arbuzov reaction of phosphites with 1-halogeno-2,3-epoxypropanes. Finally, the epoxidation of tertiary phosphine oxides by peroxy acids [53]. Phosphorus-containing flame retardants can be generally classified into three categories: (1) simple reactive phosphate monomers; (2) linear polyphosphazenes; (3) aromatic cyclic phosphazenes. They may be integrated into the ERA chains through co-polymerisation, homo-polymerisation, surface modification or blending; simple inorganic or organic additives are excluded [25, 53–55]. During the burning of ERAs, most of the current phosphorus-containing flame retardants may act simultaneously in the condensed and gaseous phases [56]. In the condensed phase, phosphorous-containing flame retardants can make the amount of carbonaceous residue or char, which acts as the thermal insulation, and a barrier of oxygen to transfer to the burning adhesive increase. Afterwards, a carbonised layer is formed. The carbonised layer prevents further pyrolysis of the corresponding ERA [57]. In the gaseous phase, some phosphorus-based additive flame retardants may produce several free radicals during the thermal decomposition process, and they may react with the free radicals which are generated from the ERA. Then, the free-radicals-supported combustion of polymers might be stopped due to the lack of fuel [58–60]. Wazarkar et al. [61] synthesised phosphorus–sulphur-containing di and tetra functional carboxyl curing agents and used them in preparing high-performance ERA and coating. The anticorrosive and flame-retardant properties of the adhesives and coatings were improved as the concentration of the flame-retardant-curing agents increased, and they exhibited excellent mechanical and chemical properties and thermal stability. Ma et al. [62] synthesised a phosphorus-containing bio-based ERA from itaconic acid and 9,10-dihydro-9-oxa-10-phosphaphenanthrene-10-oxide. As the matrix, its cured epoxy network with methyl hexahydrophthalic anhydride as the curing agent showed comparable Tg and mechanical properties to diglycidyl ether in a bisphenol A system, as well as good flame retardancy with UL94 V-0 grade during a vertical burning test.

3. Application of ERAs

ERA is often used mainly because it has the advantages of low-temperature curing, good bonding performance and improved engineering efficiency. ERAs are usually used for metal bonding, concrete bonding and wood adhesive.

3.1 Application of ERA for metals

Several studies have been conducted on the use of ERAs for metal bonding. In a previous study, polyurethane 1,2-polyethylene oxide was first cured onto the metal surface and then used 1,1-polyoxyethylene ether. The adhesive adheres the polyurethane resin material to the metal [63]. Subsequently, another study prepared a novel high-temperature curing epoxy adhesive using polysulfone as a raw material. Studies have shown that the tackifier resin accelerates the bonding of ERA and steel and promotes the bonding and vulcanisation of fluorine rubber [64]. Uehara et al. [65] prepared a monolithic ERA layer with a porous structure on the surface of a stainless steel (SUS) plate through a polymerisation-induced phase separation process, as a mediator for bonding SUS and various thermoplastic resin plates. The research results show that the bond strength of the apparent bond area between SUS and ERA is two to three times higher than those for direct metal-resin bonding.

3.2 Application of ERA for concrete bonding

Concrete is the most widely used worldwide building material. Traditional cement is used as a binding material for concrete. Cement concrete also has some shortcomings, such as low tensile and flexural strength, high porosity, low durability and abrasion resistance and longer solidification time. The mechanism of epoxy concrete is such that ERA forms a 3D structure through cross-linking in the combination of masonry mortar and concrete [66]. Afterwards, the ERA particles are dispersed into the system. Then, a part of the adhesive particles settle on the surface of the aggregate particles and participate in the cross-linking reaction. Finally, the ERA particles form a cured network structure, and the aggregates are bonded in the cured epoxy network structure. In ERA/mortar/concrete composites, the adhesive network forms a bridge between aggregates, so that the epoxy-based composites have higher mechanical properties and impermeability [67]. Also, it was found that the resin bond concrete has great advantages in the manufacture of machine tool beds. Kim et al. [68] studied ERA as the matrix material of resin concrete. The reinforced aggregate of resin concrete comprises pebbles and sand. The results showed that when the mass fraction of resin in the resin concrete is 7.5%, the thermal expansion coefficient of the resin concrete is the same as that of cast iron. In this case, the specific heat of resin concrete is 63% larger than that of cast iron. Beutel et al. [69] added epoxy resin-based adhesive concrete and coarse aggregate to the mixture and found that adding the aggregate did not affect the strength of the mixture. Additionally, the specimens exhibited much higher tensile and flexural strength than ordinary concrete, and the compressive strength was similar to ordinary high-strength concrete.

3.3 Application of ERA for wood bonding

Lei et al. [70] modified soy protein isolate with a surfactant grafted with maleic anhydride and blended with epoxy resin to prepare wood adhesives with higher bonding strength and good water resistance. Zhang et al. [71–73] prepared wood adhesive with good water and heat resistance by the co-polycondensation of tannin or lignin or starch with furfuryl alcohol and blending with epoxy resin. After bonding with wood, the adhesive was cured at high temperature under pressing to prepare a wood-based panel. After testing, the material showed good shear strength.

4. Conclusions

Although research activities on modifying ERAs have greatly progressed, several problems still need urgent resolution. For example, most of the toughening methods of ERAs are at the expense of the rigidity and strength of the modified product, and it is difficult to increase the toughness and strength of the ERA simultaneously. The ERA modified with rubber or nano-particles, due to the large specific surface area of nano-particles or rubber, is very easy to agglomerate. Therefore, how to uniformly disperse it in the ERA system to obtain a reinforced and toughened high-performance ERA remains an important research topic. Therefore, the future development direction of ERAs should be towards low-temperature fast curing, high performance, green environmental protection and multifunctional development. ERAs have developed in a more stable, safe and scientific direction.

Acknowledgements

This work was supported by The Yunnan Provincial Natural Science Foundation (Grant No. 202101AT070038, 2018FG001095), and the Yunnan Provincial Youth top talent project (YNWR-QNBJ-2020-166) and Youth talent support project and Middle-age Reserve Talents of Academic and Technical Leaders (2019HB026) and the 111 project (D21027).

Conflict of interest

The authors declared that they have no conflicts of interest.

Author details

Jun Zhang[1*], Hai Luo[2], Xiaojian Zhou[1] and Bowen Liu[1]

1 Yunnan Provincial Key Laboratory of Wood Adhesives and Glued Products, Southwest Forestry University, Kunming, China

2 East China Woody Fragrance and Flavor Engineering Research Center of National Forestry and Grassland Administration, College of Forestry, Jiangxi Agricultural University, Nanchang, China

*Address all correspondence to: zj8101274@163.com

References

[1] Shahrokhinia A, Biswas P, Reuther JF. Orthogonal synthesis and modification of polymer materials. Journal of Polymer Science. 2021;**59**(16):1748-1786

[2] Song S, Yang S, Li C, et al. Preparation of nanometer conductive powder cul and study on its performances. Journal of Donghua University, Natural Science. 2004;**30**(5):97-101

[3] Zhang YL. Epoxy Adhesive. Beijing: Chemical Industry Press; 2017. pp. 1-6

[4] Yu S, Li X, Zou M, et al. Effect of the aromatic amine curing agent structure on properties of epoxy resin syntactic foams. ACS Omega. 2020;**5**(36):23268-23275

[5] Cao J, Duan H, Zou J, et al. A bio-based phosphorus-containing co-curing agent towards excellent flame retardance and mechanical properties of epoxy resin. Polymer Degradation and Stability. 2021;**187**:109548

[6] Huang S, Kong X, Xiong Y, et al. An overview of dynamic covalent bonds in polymer material and their applications. European Polymer Journal. 2020;**141**(80):110094

[7] Levchik SV, Weil ED. Thermal decomposition, combustion and flame-retardancy of epoxy resins—A review of the recent literature. Polymer International. 2010;**53**(12):1901-1929

[8] Hodgkin JH, Simon GP, Varley RJ. Thermoplastic toughening of epoxy resins: A critical review. Polymers for Advanced Technologies. 2015;**9**(1):3-10

[9] Mai G. Failure mechanisms in toughened epoxy resins—A review. Composites Science and Technology. 1988;**31**(3):179-223

[10] Dong W, Yao SS, Wang S. Reform and practice of production practice for our college polymer material and engineering major. Polymer Bulletin. 2016;**12**:66-70

[11] Shao KC. Research on the toughening modification of epoxy resin adhesive. Rubber Technology and Equipment (Plastics). 2016;**42**:89-90

[12] Zhao JL, Zhang GC, Li HQ, et al. Research progress in toughening epoxy resin. Plastics. 2002;**31**(3):40-44

[13] Pouladvand AR, Mortezaei M, Fattahi H, et al. A novel custom-tailored epoxy prepreg formulation based on epoxy-amine dual-curable systems. Composites Part A: Applied Science and Manufacturing. 2020;**132**:105852

[14] Yin H, Li B, Liu YT, et al. Synthesis and properties of polyamide curing agent for epoxy resin. China Adhesives. 2016;**25**(1):9-12

[15] Chen L, Zhu BY, Zhang HQ, et al. Research progress of waterborne polyamide curing agent. Shanghai Paint. 2021;**59**(1):34-39

[16] Stemmelen M, Lapinte V, Habas JP, et al. Plant oil-based epoxy resins from fatty diamines and epoxidized vegetable oil. European Polymer Journal. 2015; **68**:536-545

[17] Bryan EH et al. Water absorption isotherms for cured bisphenol A type epoxy resins: Effects of isotherm temperature and resin hardener. Journal of Applied Polymer Science. 1984;**29**(6):2021-2038

[18] Gholipour-Mahmoudalilou M, Roghani-Mamaqani H, Azimi R, et al. Preparation of hyperbranched

poly (amidoamine)-grafted graphene nanolayers as a composite and curing agent for epoxy resin. Applied Surface Science. 2018;**428**(15):1061-1069

[19] Patel BP, Patel HS, Patel SR. Modified castor oil as an epoxy resin curing agent. Journal of Chemistry. 2012;**1**(1):11-16

[20] Wan J, Bu ZY, Xu CJ, et al. Preparation, curing kinetics, and properties of a novel low-volatile starlike aliphatic-polyamine curing agent for epoxy resins. Chemical Engineering Journal. 2011;**171**(1):357-367

[21] Xu YH, Yue CY, Wang WJ. Synthesis and properties of modified alicyclic amine curing agent. Adhesion. 2007;**28**(5):17-18

[22] Yin H, Jin H, Wang C, et al. Thermal, damping, and mechanical properties of thermosetting epoxy-modified asphalts. Journal of Thermal Analysis & Calorimetry. 2014;**115**:1073-1080

[23] Hao B, Liu Y, Yu X, et al. Synthesis, polymerization kinetics and thermal properties of benzoxazine resin containing ortho -maleimide functionality. Macromolecular Research. 2021;**29**(1):24-32

[24] Galià M, Serra A, Mantecón A, et al. Curing reaction of diglycidylesters containing alicyclic imide structures with anhydrides and amines as hardeners. Journal of Applied Polymer Science. 2010;**57**(4):413-420

[25] Liu D, Zhao WJ, Wu F, et al. The effect of curing agent molecular structure on the friction and corrosion properties of epoxy resin coatings. Automotive and Motorcycle Tribological Materials Advanced Technology and Application Promotion Association. 2015;**472**:85-91

[26] Freitas A. Polymer degradation and stability. Polymer Science. 2006;**87**(87):171-181

[27] Luo ZW, Yu XH, Chen Q, et al. Development of bismaleimide modified epoxy adhesive. Adhesion. 2015;**36**(12):56-59

[28] Cheng J, Yan HQ, Fu SY, et al. Preparation and thermal properties of epoxy resin containing naphthalene ring. Journal of Fujian College of Forestry. 2011;**31**(3):281-284

[29] Yang MS, Ya L, Li LK, et al. Preparation and properties of special epoxy resin containing dicyclopentadiene and naphthalene ring. Modern Plastics Processing and Application. 2010;**22**(4):5-8

[30] Zhang B, Liu WQ. Organic silicon modified epoxy resin. New Chemical Materials. 2001;**29**(8):5

[31] Hu CH, Liu WQ, Wang ZF, et al. Study on heat resistance, toughness and flame retardancy of silicone-phosphorus hybrid modified epoxy resin. China Plastics. 2009;**23**(10):73-77

[32] Ramirez FA, Carlsson LA, Acha BA. Evaluation of water degradation of vinylester and epoxy matrix composites by single fiber and composite tests. Journal of Materials Science. 2008;**43**(15):5230-5242

[33] Dong LL, Li L. Application research progress of epoxy resin in electronic industry. Fine and Specialty Chemicals. 2013;**21**(08):35-38

[34] Gui D, Gao X, Hao J, et al. Preparation and characterization of liquid crystalline polyurethane-imide modified epoxy resin composites. Polymer Engineering & Science. 2014;**54**(7):1704-1711

[35] Wang HX, Zheng YQ, Hou YB, et al. Development of one-component epoxy resin structural adhesive with high shear and high peel strength. China Adhesives. 2008;**17**(8):19-12

[36] Lou CH, Liu XJ, et al. Functional dendritic curing agent for epoxy resin: Processing, mechanical performance and curing/toughening mechanism. Composites Part B: Engineering. 2018;**136**:20-27

[37] Zheng YP, Ning RC. Study on properties of nano-SiO_2 epoxy resin composites. FRP/Composite Materials. 2001;**1**(2):34-36

[38] Chen L, Wang YZ. A review on flame retardant technology in China. Part 1: Development of flame retardants. Polymers for Advanced Technologies. 2010;**21**(1):1-26

[39] Qian L, Qiu Y, Wang J, et al. High-performance flame retardancy by char-cage hindering and free radical quenching effects in epoxy thermosets. Polymer. 2015;**68**:262-269

[40] Tang S, Wachtendorf V, Klack P, et al. Enhanced flame-retardant effect of a montmorillonite/phosphaphenanthrene compound in an epoxy thermoset. RSC Advances. 2017;**7**(2):720-728

[41] Katsoulis C, Kandola BK, Myler P, et al. Post-fire flexural performance of epoxy-nanocomposite matrix glass fibre composites containing conventional flame retardants. Composites Part A-Applied Science and Manufacturing. 2012;**43**(8):1389-1399

[42] Petrie EM. Epoxy Adhesive Formulations. New York: McGraw-Hill; 2006

[43] Le W, Deng C, Zhao ZY, et al. Flame retardation of natural rubber: Strategy and recent progress. Polymers. 2020;**12**(2):429

[44] Morgan AB. The future of flame retardant polymers—Unmet needs and likely new approaches. Polymer Reviews. 2019;**59**(1):25-54

[45] Liu Q, Wang D, Li Z, et al. Recent developments in the flame-retardant system of epoxy resin. Materials. 2020;**13**(9):2145

[46] Beach MW, Hull JW, King BA, et al. Development of a new class of brominated polymeric flame retardants based on copolymers of styrene and polybutadiene. Polymer Degradation and Stability. 2017;**135**:99-110

[47] Jiang Y, Han X, Gong J, et al. Laboratory investigation of epoxy asphalt binder modified by brominated SBS. Construction and Building Materials. 2019;**228**:116733

[48] Wu Z, Li S, Liu M, et al. Study on liquid oxygen compatibility of bromine-containing epoxy resins for the application in liquid oxygen tank. Polymers for Advanced Technologies. 2016;**27**(1):98-108

[49] Cheng J, Wang J, Yang S, et al. Benzimidazolyl-substituted cyclotriphosphazene derivative as latent flame-retardant curing agent for one-component epoxy resin system with excellent comprehensive performance. Composites Part B-Engineering. 2019;**177**:107440

[50] Wu Q, Lv J, Qu B. Preparation and characterization of microcapsulated red phosphorus and its flame-retardant mechanism in halogen-free flame retardant polyolefins. Polymer International. 2003;**52**(8):1326-1331

[51] Horold S. Phosphorus flame retardants in thermoset resins.

Polymer Degradation and Stability. 1999;**64**(3):427-431

[52] May CA. Epoxy Resins: Chemistry and Technology. New York: Taylor & Francis Group; 1988

[53] Troitzsch J. Flame retardant polymers current status and future trends. Makromolekulare Chemie. Macromolecular Symposia. 1993;**74**(1):125-135

[54] Rad ER, Vahabi H, Anda A, et al. Bio-epoxy resins with inherent flame retardancy. Progress in Organic Coatings. 2019;**135**:608-612

[55] Velencoso MM, Battig A, Markwart JC, et al. Molecular firefighting-how modern phosphorus chemistry can help solve the challenge of flame retardancy. Angewandte Chemie. 2018;**57**(33):10450-10467

[56] Veen I, Boer JD. Phosphorus flame retardants: Properties, production, environmental occurrence, toxicity and analysis. Chemosphere. 2012;**88**(10):1119-1153

[57] Lligadas G, Ronda JC, Galla M, et al. Synthesis and properties of thermosetting polymers from a phosphorous-containing fatty acid derivative. Journal of Polymer Science Part A: Polymer Chemistry. 2006;**44**(19):5630-5644

[58] Dong LP, Deng C, Li RM, et al. Poly (piperazinyl phosphamide): A novel highly-efficient charring agent for an EVA/APP intumescent flame retardant system. RSC Advances. 2016;**6**(36):30436-30444

[59] Yang AH, Deng C, Chen H, et al. A novel Schiff-base polyphosphate ester: Highly-efficient flame retardant for polyurethane elastomer. Polymer Degradation and Stability. 2017;**144**:70-82

[60] Yang J, Zhao Y, Li M, et al. A review of a class of emerging contaminants: The classification, distribution, intensity of consumption, synthesis routes, environmental effects and expectation of pollution abatement to organophosphate flame retardants (OPFRs). International Journal of Molecular Sciences. 2019;**20**(12):2874

[61] Wazarkar K, Sabnis A. Synergistic effect of P S and crosslink density on performance properties of epoxy coatings cured with cardanol based multifunctional carboxyl curing agents. Reactive & Functional Polymers. 2018;**128**:74-83

[62] Ma S, Liu X, Jiang Y, et al. Synthesis and properties of phosphorus-containing bio-based epoxy resin from itaconic acid. Science China Chemistry. 2014;**57**(3):379-388

[63] Lee MC, Ho TH, Wang CS. Synthesis of tetrafunctional epoxy resins and their modification with polydimethylsiloxane for electronic application. Journal of Applied Polymer Science. 1996; **62**(1):217-225

[64] Lin ST, Huang SK. Thermal degradation study of siloxane-DGEBA epoxy copolymers. European Polymer Journal. 1997;**33**(3):365-373

[65] Uehara F, Matsumoto A. Metal-resin bonding mediated by epoxy monolith layer. Applied Adhesion Science. 2016;**4**(1):18

[66] Zhao C, Zhang G, Zhao L. Effect of curing agent and temperature on the rheological behavior of epoxy resin systems. Molecules. 2012;**17**(7):8587-8594

[67] Jiao C, Dong J, Zhang C, et al. Synthesis and properties of a phosphate

ester as curing agent in an epoxy resin system. Iranian Polymer Journal. 2014;**23**(8):591-598

[68] Kim HS, Park KY, Dai GL. A study on the epoxy resin concrete for the ultra-precision machine tool bed. Journal of Materials Processing Technology. 1995;**48**(1-4):649-655

[69] Beutel A. Optimal Mix Design for Epoxy Resin Polymer Concrete. Australia: University of Southern Queensland; 2015

[70] Lei W, Yang T, Wang KJ, et al. Preparation and characterization of epoxy resin modified soybean-based wood adhesive. Soybean Science. 2010;**29**(1):118-120

[71] Zhang J, Liu BW, Zhou YX. Preparation of a starch-based adhesive cross-linked with furfural, furfuryl alcohol and epoxy resin. International Journal of Adhesion and Adhesives. 2021;**110**:102958

[72] Zhang J, Xi XD, Liang JK. Tannin-based adhesive cross-linked by furfuryl alcohol-glyoxal and epoxy resins. International Journal of Adhesion and Adhesives. 2019;**94**:47-52

[73] Zhang J, Wang WL, Zhou XJ. Lignin-based adhesive crosslinked by furfuryl alcohol–glyoxal and epoxy resins. Nordic Pulp & Paper Research Journal. 2019;**34**(2):228-238

Chapter 6

Epoxy Composites for Radiation Shielding

Hayriye Hale Aygün

Abstract

Due to the increase in use of radiation energy in many industrial applications, radiation shielding has become a crucial topic in order to diminish its hazardous effects. Radiation shields can be of various weights depending on the materials from which they are produced and the area in which they are used. In this sense, polymer composites have taken attention by researchers because it is aimed to obtain shields with good processability, sufficient flexibility, low weight, and subsequent performance properties. Epoxy resin is one of the mostly used synthetic polymers as a matrix element in composite material production due to its improving characteristics by means of electrical insulation, chemical resistance, service life, bonding characteristic, and mechanical properties. Besides, epoxies have intermediate radiation shielding characteristics as well. By loading epoxy matrix with fibers and/or fillers having different radiation absorption rates or mechanical resistance properties, multifunctional shields can be produced to serve in numerous applications. This chapter focuses on radiation shielding efficiency of fiber-reinforced epoxy composites and the role of fillers and fiber-based materials on manufacturing of functional radiation shields.

Keywords: composite, epoxy resin, fiber, filler, radiation, shielding

1. Introduction

For protecting humans and the environment from the hazardous effects of radiation, various forms of shields have been used in different fields in which radiation has been utilized. Shielding materials manufactured from lead, stainless steel, and concrete are heavy and rigid structures and also not resistant to corrosion. These structures have been generally used as blocks for shielding against radiation and have not sufficient flexibility and comfort properties in order to be used in shielding garments (**Figure 1**). Therefore, researchers have been focused on the manufacturing of advanced materials with good shielding capability, lightweight, high modulus, and mechanical properties. At this point, composite materials have taken the attention of researchers due to manufacturing of a unique material from different materials with dissimilar characteristics.

Epoxy polymer is one of the most important thermoset polymers used in composite manufacturing due to its good wetting ability, low cure shrinkage, excellent chemical corrosion resistance, good dimensional stability, high tensile, fatigue, and compression strength. By courtesy of its use in composites, it contributes to

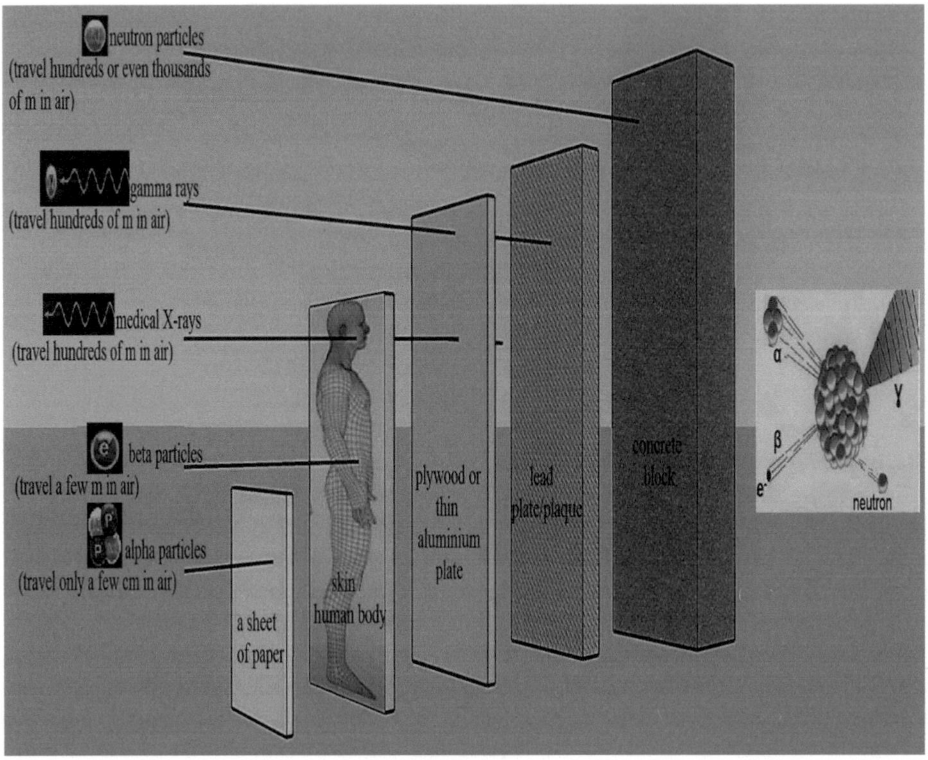

Figure 1.
Types and penetration depths of radiation.

the properties of the whole composite by means of good mechanical strength, high stiffness, excellent chemical resistance, flame retardancy, and high electrical strength [1, 2]. Solid epoxy polymer is the output material obtained by reaction of curing agent and liquid resin. There are various types of epoxy-based liquid resin because the numbers of epoxide groups on its starting material can be variable. Diglycidyl ether of bisphenol A (DGEBA) has two epoxide groups in its structure, which is the most common starting material used in the manufacturing of epoxy-based liquid resin. For solidification, liquid resin is treated with small amounts of reactive curing agent and then a tridimensional network occurs as a result of crosslinking. The use of different types of starting materials and reactive curing agents results in various types of solid epoxy polymers with different characteristics. Thereby, the properties of epoxy resin are given a range of values as it is seen in **Table 1**. In case of being cured, the system exhibits brittle characteristic due to crosslinking mentioned above, and this case causes incomparable decreases in their relevant mechanical properties, especially in impact strength [3]. In addition to crosslinking occurred by reaction between the curing agent and epoxy resin, some additional structural changes are observed in case of irradiation of epoxy-based system. The color of epoxy resin alters from transparent to yellow and resin can be even degraded depending on exposing dose and starting material used for manufacturing epoxy resin. Even so, epoxies are addressed as assuring matrix elements with high radiation stability for composite manufacturing [4]. In order to

Property	Value
Density (gcm^{-3})	1.2–1.3
Tensile strength (MPa)	55–130
Tensile modulus (GPa)	2.75–4.10
Poisson's ratio	0.2–0.33
Thermal expansion coefficient	50–80
Cure shrinkage (%)	1–5

Table 1.
Typical properties of epoxy resin at room temperature.

limit brittle characteristics and develop mechanical performance of epoxy-based systems, epoxy resins should be reinforced with flexible materials.

High strength-to-weight ratio is one of the advantages of fiber-reinforced composites. By increasing interactions between fiber and epoxy matrix, the resistance of the whole composite to many destructive forces is improved. The improvement can be successfully achieved by the incorporation of elastomeric/thermoplastic phases or by adding organic/inorganic particles into epoxy resins [5–7]. Gojny et al. dispersed carbon-based nanoparticles in epoxy resins and reported that fracture toughness of produced composites effectively improved at low nanoparticle concentration as well as stiffness [8]. In another study, carbon-based nanoparticles lead to the development of flexural strength and modulus of the epoxy composite [9]. There are numerous researches about the use of inorganic and organic particles in reinforcement of epoxies and improvement of composite properties in different ways [10–20].

2. Polymers and fillers used for radiation shielding

Polymers have been intensively used for fabricating radiation shields due to their lightness, low cost, and elasticity [21, 22]. However, polymers behave differently when they are irradiated. Under different radiation sources with variable frequency rays, crosslinking, chain scission or degradation may be observed in a polymer chain [23]. The behaviors of some irradiated polymers and classifications according to their radiation resistance are given in **Table 2**.

In order to delay the degradation of polymeric structures and diminish the hazardous effect of radiation on polymers, fillers are added to the structure during manufacturing process. The radiation shielding efficiency of a filler depends on its atomic number and the atomic structure of filler is a crucial factor in order to fabricate functional structures from polymer and filler for intended end-use. Fillers with high atomic number are generally used for gamma radiation shielding (**Table 3**). However, the use of fillers with low atomic number is preferable for neutron radiation shielding. When considering that there are generally low atomic number elements in a polymer chain, compatibility of polymer/filler combination has a significant effect on radiation shielding efficiency of fabricated composite material [25]. Many researches have been performed on the use of polymers [26–30] and fillers [31–33] for the manufacturing of radiation shields.

	Polymer	Reaction when irradiated	Exposure dose	
			Neutron (neutron/cm^2)	Gamma (R)
Polymers with very low radiation resistance	PTFE	Chain scission	10^{13}	10^5–10^6
	PMMA	Chain scission	10^{14}	10^6
	Butyl rubber	Chain scission	—	10^6
Polymers with low radiation resistance	PVC	Chain scission or crosslinking and degradation	10^{14}	$5*10^7$–10^8
	Cellulose acetate	Chain scission	10^{14} –10^{15}	$5*10^6$ – $4*10^7$
	Phenol formaldehyde resins	—	$7*10^{14}$	10^7
	PA	Crosslinking	$4*10^{14}$	10^7–$5*10^7$
	Butadiene styrene rubber	Crosslinking	10^{15}–10^{16}	10^7
Polymers with medium radiation resistance	PE	Crosslinking	10^{17}	10^8
	PET	Chain scission or crosslinking and degradation	10^{17}–10^{18}	10^8–$2*10^8$
	Epoxy	Crosslinking	10^{17}	$5*10^8$
Polymers with high radiation resistance	Polystyrene	Crosslinking	10^{18}–10^{19}	$5*10^9$

Table 2.
Reactions and resistance of polymers against radiation [23, 24].

Element	Atomic number	Density(gcm^{-3})	Absorption edge (keV)
Cadmium (Cd)	48	8.65	26.7
Tin (Sn)	50	7.30	29.2
Antimony (Sb)	51	6.69	30.5
Cesium (Cs)	55	1.87	26.0
Barium (Ba)	56	3.50	37.4
Gadolinium (Gd)	64	7.90	50.2
Tungsten (W)	74	19.30	69.5
Lead (Pb)	82	11.36	88.0
Bismuth (Bi)	83	9.75	90.5

Table 3.
Comparison of some elements used in radiation shielding [34].

3. Filler loaded epoxy composites for radiation shielding

There are many attempts for improving the shielding characteristics of epoxy or other polymer matrix with elemental particles. The researchers observed effects of particle loading via numerous trials by means of minimizing particle size, doping matrix with different particle concentration in weight, exposing composite specimens to different radiation sources, analyzing microstructural changes under different radiation doses, and testing failure mechanism of composites. Radiation shielding properties of epoxy-based composite panels were tested by Al-Sarray et al. Epoxy resin was loaded with different barite concentrations of 0–50 wt %, and the linear attenuation coefficient of fabricated composites was tested by Co^{60} and Cs^{137} radioactive sources. Radiation shielding efficiency improved with the increase in barite concentration [35]. Ergin et al. compared the effects of lead oxide and barium oxide on the radiation shielding performance of epoxy composites. They reported that gamma radiation performance handled by lead oxide/epoxy composites could be obtained by the addition of barium oxide but the weight percentage of barium oxide must be two times more than lead oxide addition. Besides, addition of barium oxide at 40 wt % in epoxy resin exhibited better radiation shielding performance than gadolinium oxide/epoxy composites, concrete, and steel [36]. Li et al. dispersed micro- and nano-gadolinium oxide particles into epoxy matrix and evaluated both mechanical properties and radiation shielding characteristics of fabricated composites. Gadolinium oxide addition enhanced shielding characteristics due to dominating photoelectric effect of the gadolinium element. Nano-gadolinium oxide/epoxy composites exhibited better X-ray and gamma shielding characteristics and had similar flexural strength but higher flexural modulus with those of micro-gadolinium oxide/epoxy composites [37]. More et al. determined radiation shielding parameters of metal chloride/epoxy composites. Doping epoxy resin with higher weight ratios of metal chloride lead to increase in attenuation parameters of epoxy composite and test results were comparable with those of pure lead metal [38].

Mechanical properties, structural characteristics, and gamma shielding efficiency of epoxy composites were reported by Alavian et al. Epoxy resin was doped with inorganic nanoparticles such as lead, zinc, zinc oxide, titanium, and titanium oxide. Increasing nanoparticle loading enhanced shielding efficiency. 25 wt% Pb/epoxy nanocomposites showed better shielding properties but low mechanical strength than those of their counterparts [39]. Degradation of epoxy composite by high-frequency rays was investigated by Saiyad et al. They fabricated three different composite materials by loading graphite, boron nitride, and lead into epoxy resin. They irradiated epoxy composites by Am-Be neutron sources. They reported that linear absorption coefficients of composites were strictly dependent on the dispersion of filler and the highest shielding performance was observed in graphite/epoxy composites [40]. Aldhuhaibat et al. examined gamma radiation shielding performance of pure epoxy resin and epoxy-based nanocomposites with aluminum oxide or ferrium oxide nanopowder at different concentrations of 10–15 wt %. Specimens were irradiated by Cs^{137}(1.05 kBq with single gamma-ray emission energy of 0.662 MeV) and Co^{60} (74 kBq with two gamma-ray emission energies of 1.173 and 1.333 MeV) radioactive sources and linear attenuation coefficients of specimens were measured by NaI detector. They claimed that epoxy nanocomposites were potential gamma radiation shields with improved characteristics [41]. Another study was performed by Azman et al. Nano-sized tungsten oxide/epoxy composites had higher attenuation properties at 22–35 kV X-ray tube voltages used in mammography and radiography units. The particle size of tungsten was found as negligible by means of transmitted beam intensity at 40–120 kV tube voltages [42]. Cheng et al. studied the radiation degradation mechanism of tungsten/epoxy composites. They tested composites specimens under different activities of Co^{60} sources. Loading with tungsten improved shielding characteristic of composites. However, measurements showed that an increase in radiation dose caused a decrease, then a slight increase and a sharp decrease in thermal and mechanical properties of composites [43].

4. Modification of epoxy resin with fibers/fillers for radiation shielding

In recent years, studies performed on radiation shields have focused on developing failure mechanisms and decreasing the weight of shield for supplying personal comfort. For this goal, researchers offered dissimilar alternative shields by studying various polymeric matrices with different fillers. Kim observed the effect of particle size and dispersibility of tungsten particles on radiation shielding performance of samples. For this aim, three types of tungsten-loaded HDPE shields were manufactured with identical thickness and sizes by doping nanoparticles, microparticles, and their mixture. It was claimed that sufficient protection was handled against low dose exposure notwithstanding particle size. But nanoparticle loaded HDPE sheets were more resistant to high-energy radiation. The shielding sheets produced with a mixture of different particle sizes of tungsten showed similar shielding performance as microparticle tungsten-loaded sheets [44]. Manufacturing and radiation shielding properties of nano gadolinium oxide/PMMA composites were searched by Shreef and Abdulzahara. Composite shields were fabricated with varying concentrations of gadolinium oxide (10–40 wt %). The thickness of composites was measured and shielding performance was tested with Co^{60} and Cs^{137} radiation sources. Test results showed that increasing nanoparticle concentration lead to an increase in thickness and improvement in attenuation coefficient but a decrease in half-value layer values of epoxy composites [45]. Zheng et al. fabricated S-glass fabric/epoxy composites with a ratio of 1:1. Composites were irradiated by Co^{60} source and the effects of irradiation on properties of composites were compared. They claimed that gamma-ray irradiation caused negligible damage on S-glass fiber and possible degradation on epoxy resin. By increasing the exposing dose of gamma radiation, the color of composite altered from yellow to brown and tensile strength of composite reduced gradually. However, composite preserved its thermal and dimensional stability and exhibited excellent thermal conductivity after irradiation [46]. Li et al. tried to produce a novel radiation shielding composite with high mechanical strength. For this aim, they fabricated erbium oxide-loaded basalt fiber/epoxy composites by prepreg autoclave process and test shielding performance of composites by exposing them to X and gamma rays. They claimed that basalt fiber/erbium oxide/epoxy composites had high mass attenuation coefficient than aluminum at low photon energies ranging from 31 keV to 80 keV [47]. The fracture toughness of carbon fabric/epoxy composites was investigated by Phong et al. They produced micro/nano-sized bamboo fibrils and dispersed these fibrils into epoxy matrix. They reported that matrix cracking was delayed, crack growing was reduced, and fracture toughness of composites was improved [48]. Haque et al. reinforced epoxy matrix with layered nano silicate particles at very low concentrations (1 wt %) and claimed that flexural strength, toughness, decomposition temperature, and interlaminar shear strength of S2-glass/epoxy composites were improved due to enhanced fiber/matrix adhesion and reduced residual stresses [49]. Recycled PET fibers were used to mimic marble material. Nguyen et al. modified calcium carbonate particles with stearic acid in order to enhance compatibility between epoxy resin and calcium carbonate. They fabricated composites by positioning a single layer of recycled PET fiber mat in the core and by coating the front and backside of PET mat with epoxy/calcium carbonate mixture. They concluded that flexural properties, impact resistance, and thermal stability of epoxy composites were improved [50]. Saleem et al. presented an empirical approach and compared radiation shielding of lead nanoparticles loaded epoxy composites with glass or carbon fiber. The results showed that lead nanoparticles improved shielding characteristics and lead to an increase in mass attenuation coefficients of composites. Mass attenuation coefficients were 0.2145 cm^2/g and 0.2152 cm^2/g for carbon and glass fiber reinforced epoxy composites at lead nanoparticle concentration of 50 wt%. But half-value length of epoxy composite with glass fiber was reported as 1.431 cm, which was lower than epoxy composites with carbon fiber (1.756 cm) [51].

Effects of radiation on neat epoxy resin and carbon fiber/epoxy composites were studied by Hoffman and Skidmore (2009). Front and back surfaces of plain woven carbon fabrics were treated with epoxy/hardener mixture (2:1). Prepared composites were exposed to mechanical and thermal testing and also analyzed by means of microstructural properties and radiation characteristics. After being irradiated, there were no remarkable changes in mechanical resistance of composites but significant variations were observed in thermal properties, spectroscopic analysis, and hardness value of neat epoxy samples as a result of gamma radiation [52]. Zhong et al. examined the cosmic radiation shielding properties of hot-pressed UHMWPE/nano-epoxy composites and they concluded that epoxy composites with the combination of continuous fibers such as UHMWPE and/or graphite nanofibers found as multifunctional hybrid systems by means of good structural properties, cost-effectiveness, and high radiation shielding performance [53]. In another study, UHMWPE/epoxy composites were fabricated by Mani et al. and test results showed that composites containing gadolinium and boron nanoparticles had good neutron shielding performance [54]. Condruz et al. suggested coating carbon/epoxy composites with different types of functional materials such as tantalum foil, babbitt and Monel for protecting hazardous effects of proton radiation. They impregnated 2 × 2 twill woven carbon fabric into epoxy resin and then coated polymeric substrates with zinc, Babbitt or zinc/Monel particles by thermal spray technique. They concluded that the coating process reduced penetration depth of ion beam and produced composites were lightweight shields for proton radiation [55]. The effects of hybridization on mechanical, thermal and radiation shielding efficiency of composites were also examined and reported by Zagaoui et al. They blended epoxy resin (90 wt%) with benzoxazine resin (10 wt%) and reinforced bicomponent matrix with silane-treated glass and basalt fibers. Hybridization of different types of resins developed mechanical and thermal properties and excellent shielding characteristic was gained by integrating hybrid fibers into bicomponent matrix system [56].

5. Conclusion

Heavy concretes, lead plates, and stainless steel blocks or plaques are known as conventional radiation shielding materials but they are heavy and not suitable for individual protective equipment. Polymers are functional lightweight materials but do not supply adequate protection on their own. Thereby advanced radiation shields should be fabricated by the composition of polymer-based materials and substances with high radiation shielding activity. At this point, material selection has crucial importance on the efficiency of protection by which radiation source it is irradiated.

The destructive effect of radiation on the material is related to the type of radiation source, exposing dose rate, exposure period, radiation absorption rate of material and strength of interbonding forces between components if a composite material is used. By taking into consideration the advantages introduced with composite manufacturing, the destructive effect of radiation can be limited by a combination of materials with high attenuation rates. At this point, the shielding efficiency of composite material depends on how components in a composite behave in case of irradiation. Epoxies, the mostly used matrix elements in composite manufacturing, exhibit physical changes such as color transition and low shrinkage percentage and mechanical changes such as a decrease in flexural and impact strength due to crosslinking when they are irradiated. Despite these changes, they are known as reliable materials for being used as matrix elements in radiation shielding. Undesired physical and mechanical changes observed in irradiated epoxies tried to be eliminated by fiber and/or filler loading for handling effective radiation shields with long life. Fiber addition into epoxy matrix causes an increase in hardness, fracture toughness, impact resistance, flexural strength and modulus, and also consistency in dimensional stability and thermal properties with respect to neat epoxy. However, loading fillers into epoxy matrix outputs composite material with inconsistent mechanical and thermal properties, especially in heterogenic filler dispersion and inappropriate particle size. Modification of epoxy with fillers having high radiation absorption rate develops radiation shielding efficiency of the whole composite but not mechanical or thermal characteristics for long-term use of composite. Thereby epoxy-based composites, which are designed to be used for radiation shields, should contain both fillers and fibers. Epoxy-based radiation shields serve as effective protectors in the case of reinforcing with fiber-based structures and fillers having high radiation absorption rate and photoelectric properties.

Fiber and filler reinforced epoxy composites are functional engineering materials and compete with some conventional radiation shields by means of strength and modulus properties per unit weight. The functionality is improved by proper fiber/matrix combination, high interfacial bonding between these constituents, functional additive/filler loading, modification of fiber surface with an appropriate sizing agent, well-dispersed nano-sized filler addition, and suitable manufacturing technique. In this way, epoxy composite serves as a unique shielding material for which goal it is fabricated and in which field it is intended to be used. Moreover, there is a need to figure out the best alternative to be used in medical diagnostics and nuclear industry.

Author details

Hayriye Hale Aygün
University of Kahramanmaras Sutcu Imam, K. Maras, Turkey

*Address all correspondence to: hhalesolak@hotmail.com

References

[1] Ratna D. Handbook of Thermoset Resins. UK: iSmithers; 2009. p. 410

[2] Aygün B. Epoxy based metal and metal oxide doped new composite neutron and gamma radiation moderator material. Erzincan University Journal of Science and Technology. 2019;**12**(3):1442-1453

[3] Mallick PK. Fiber Reinforced Composites: Materials, Manufacturing and Design. New York: CRC Press; 2007. p. 638

[4] Akbari R, Beheshty MH, Shervin M. Toughening of dicyandiamide-cured DGEBA based epoxy resins by CTBN liquid rubber. Iranian Polymer Journal. 2013;**22**:313-324

[5] Özdemir T, Usanmaz A. Degradation of poly(bisphenol- a-epichlorohydrin) by gamma irradiation. Radiation Physics and Chemistry. 2008;**77**:799-805

[6] Njuguna J, Pielichowski K, Alcock JR. Epoxy-based fibre reinforced nanocomposites. Advanced Engineering Materials. 2007;**9**(10):835-847

[7] Dodiuk H, Kenig S, Blinsky I, Dotan A, Buchman A. Nanotailoring of epoxy adhesives by polyhedral-oligomeric-sil-sesquioxanes (POOS). International Journal of Adhesion and Adhesives. 2005;**25**:211-218

[8] Gojny FH, Wichmann MHG, Köpke U, Fiedler B, Schulte K. Carbon nanotube reinforced epoxy composites: Enhanced stiffness and fracture toughness at low nanotube content. Composites Science and Technology. 2004;**64**:2363-2371

[9] Kong J, Ning R, Tang Y. Study on modification of epoxy resins with acrylate liquid rubber containing pendant epoxy groups. Journal of Materials Science. 2006;**41**:1639-1641

[10] Zhao DL, Qiao RH, Wang CZ, Shen ZM. Microstructure and mechanical property of carbon nanotube and continuous carbon fiber reinforced epoxy resin matrix composites. Advanced Materials Research. 2006;**11**:517-520

[11] Chisholm N, Mahfuz H, Rangari VK, Ashfaq A, Jeelani S. Fabrication and mechanical characterization of carbon/SiC-epoxy nanocomposites. Composite Structures. 2005;**67**(1):115-124

[12] Schmidt H. New type of non-crystalline solids between inorganic and organic materials. Journal of Non-Crystalline Solids. 1985;**73**:681-691

[13] Wang K, Chen L, Wu J, Toh ML, He C, Yee AF. Epoxy nanocomposites with highly exfoliated clay: Mechanical properties and fracture mechanisms. Macromolecules. 2005;**38**(3):788-800

[14] Mark JE. Ceramic reinforced polymers and polymer modified ceramics. Polymer Engineering and Science. 1996;**36**(24):2905-2920

[15] Gilbert EN, Hayes BS, Seferis JC. Variable density composite systems constructed by metal particle modified prepregs. Journal of Composite Materials. 2002;**36**(17):2045-2060

[16] Timmerman JF, Hayes BS, Seferis JC. Nanoclay reinforced effects on the cryogenic microcracking of carbon fiber/epoxy composites. Composites Science and Technology. 2002;**62**(9):1249-1258

[17] Brunner AJ, Necola A, Rees M, Gasser P, Kornmann X, Thomann R, et al.

The influence of silicate-based nano-filler on the fracture toughness of epoxy resin. Engineering Fracture Mechanics. 2006;**73**(16):2336-2345

[18] Mohan RV, Kelkar AD, Akinyede O. Vartm processing and characterization of composite laminates from epoxy resins dispersed with alumina particles. In: Proceedings of 50th International SAMPE Symposium and Exhibition; 1-5 May 2005; USA. Long Beach: SAMPE; 2005. pp. 2425-2431

[19] Kornmann X, Rees M, Thomann Y, Necola A, Barbezat M, Thomann R. Epoxy-layered silicate nanocomposites as matrix in glass fibre-reinforced composites. Composites Science and Technology. 2005;**65**(14):2259-2268

[20] Ragosta G, Abbate M, Musto P, Scarinzi G, Mascia L. Epoxy-silica particulate nanocomposites: Chamical interactions, reinforcement and fracture toughness. Polymer. 2005;**46**(23):10506-10516

[21] Kaphle A, Umapathi NPNA, Daima H. Nanomaterials for agriculture, food and environment: Applications, toxicitysss and regulation. Environmental Chemistry Letters. 2018;**16**:43-58

[22] More CV, Alsayed Z, Badaw MS, Thabet AA, Pawar PP. Polymeric composite materials for radiation shielding: A review. Environmental Chemistry Letters. 2021;**19**:2057-2090

[23] Ivanov VS. Radiation Chemistry of Polymers. Wakefield: VSP Publishing; 1992. p. 321

[24] Aygün HH. Production and Characterization of Shielding Surfaces with Bismuth Containing Polymeric Materials against X-Ray Radiation [PhD Thesis]. Kahramanmaras: University of Kahramanmaras Sutcu Imam; 2018

[25] Kaçal M, Akman F, Sayyed M. Evaluation of gamma-ray and neutron attenuation properties of some polymers. Nuclear Engineering and Technology. 2019;**51**(3):818-824

[26] Kilicoglu O, Kara U, Inanc I. The impact of polymer additive for N95 masks on gamma-ray attenuation properties. Materials Chemistry and Physics. 2021;**260**(16):124093

[27] Mirji R, Lobo B. Computation of the mass attenuation coeffcient of polymeric materials at specifc gamma photon energies. Radiation Physics and Chemistry. 2017;**135**:32-44

[28] Sayyed MI. Investigation of shielding parameters for smart polymers. Chinese Journal of Physics. 2016;**54**(3):408-415

[29] Bhosale RR, More CV, Gaikwad DK, Pawar PP, Rode MN. Radiation shielding and gamma ray attenuation properties of some polymers. Nuclear Technology and Radiation Protection. 2017;**32**(3):288-293

[30] Mann K, Rani A, Heer M. Shielding behaviors of some polymer and plastic materials for gamma-rays. Radiation Physics and Chemistry. 2015;**106**:247-254

[31] Kaçal MR, Dilsiz K, Akman F, Polat H. Analysis of radiation attenuation properties for polyester/Li_2WO_4 composites. Radiation Physics and Chemistry. 2021;**179**:109257

[32] Lanina S, Kaminskaya N, Benyaev N, Suslova V, Grigorevskaya M. On possible use of inorganic fillers and matrix polymers in radiation shielding materials. Biomedical Engineering. 2013;**46**(6):228-231

[33] Engelmann HJ. Material for shielding from radiation; 2009. WO2009097833A1

[34] McCaffrey JP, Shen H, Downtown B, Mainegra-Hing E. Radiation attenuation by lead and nonlead materials used in radiation shielding garments. Medical Physics. 2007;**34**(2):530-537

[35] Al-Sarray E, Günoğlu K, Evcin A, Bezir NÇ. Radiation shielding properties of some composite panel. Acta Physica Polonica A. 2017;**132**:490-492

[36] Ergin Y, Karabul Y, Guven Ozdemir Z, Kılıç M. Experimental comparison of PbO and BaO addition effect on gamma ray shielding performance of epoxy polymer. European Journal of Science and Technology. 2019;**16**:256-266

[37] Li R, Gu Y, Wang Y, Yang Z, Li M, Zhang Z. Effect of particle size on gamma radiation shielding property of gadolinium oxide dispersed epoxy resin matrix composite. Materials Research Express. 2017;**4**(3):035035

[38] More CV, Pawar PP, Badawi MS, Thabet AA. Extensive theoretical study of gamma-ray shielding parameters using epoxy resin-metal chloride mixtures. Nuclear Technology and Radiation Protection. 2020;**35**(2):138-149

[39] Alavian H, Samie A, Tavakoli-Anbaran H. Experimental and Monte Carlo investigations of gamma ray transmission and buildup factors for inorganic nanoparticle/epoxy composites. Radiation Physics and Chemistry. 2020;**174**:108960

[40] Saiyad DM, Devashrayee N, Mewada R. Study the efect of dispersion of filler in polymer composite for radiation shielding. Polymer Composites. 2014;**35**(7):1263-1266

[41] Al-Dhuhaibat M, Salman M, Jubier N, Salim A. Improved gamma radiation shielding traits of epoxy composites: Evaluation of mass attenuation coefficient, effective atomic and electron number. Radiation Physics and Chemistry. 2020;**179**:109183

[42] Azman NN, Siddiqui S, Hart R, Low IM. Effect of particle size, filler loadings and X-ray tube voltage on the transmitted X-ray transmission in tungsten oxide-epoxy composites. Applied Radiation and Isotopes. 2013;**71**(1):62-67

[43] Chang L, Zhang Y, Liu Y, Fang J, Luan W, Yang X, et al. Preparation and characterization of tungsten/epoxy composites for γ-rays radiation shielding. Nuclear Instruments and Methods Physics Research Section B: Beam Interactions with Materials and Atoms. 2015;**356-357**:88-93

[44] Kim SC. Analysis of shielding performance of radiation shielding materials according to particle size and clustering effects. Applied Sciences. 2021;**11**:4010

[45] Sheerif AM, Abdulzahara NA. Manufacture of shielding for attenuation ionization ray by preparation of nano-gadolinium oxide with PMMA. Neuroquantology: An Interdiciplineray Journal of Neuroscience and Quantum Physics. 2021;**19**(8):66-78

[46] Zheng LF, Wang LN, Wang ZZ, Wang L. Effects of γ-ray irradiation on the fatigue strength, thermal conductivities and thermal stabilities of the glass fibres/epoxy resins composites. Acta Metallurgica Sinica. 2018;**1**(31):105-112

[47] Li R, Gu Y, Zhang G, Yang Z, Li M, Zang Z. Radiation shielding property of structural polymer composite: Continuous basalt fiber reinforced epoxy matrix composite containing

erbium oxide. Composites Science and Technology. 2017;**143**:67-74

[48] Phong NT, Gabr MH, Okubo K, Chuong B, Fujii T. Enhancement of mechanical properties of carbon fabric/epoxy composites using micro/nano-sized bamboo fibrils. Materials and Design. 2013;**47**:624-632

[49] Haque A, Shamsuzzoha M, Hussain F, Dean D. S2-glass/epoxy polymer nanocomposites: Manufacturing, structures, thermal and mechanical properties. Journal of Composite Materials. 2003;**37**(20):1821-1837

[50] Nguyen M, Vu T, Nguyen T, Nguyen T, Ha Thuc N, Bui QB, et al. Synergistic infuences of stearic acid coating and recycled PET microfbers on the enhanced properties of composite materials. Materials. 2020;**13**(6):1461

[51] Saleem RAA, Abdelal N, Alsabbagh A, Al-Jarrah M, Al-Jawarneh F. Radiation shielding of fiber reinforced polymer composites incorporating lead nanoparticles: An empirical approach. Polymers. 2021;**13**(21):3699

[52] Hoffman AN, Skidmore TE. Radiation effects on epoxy carbon fiber composite. Journal of Nuclear Materials. 2009;**392**(2):371-378

[53] Zhong W, Sui G, Jana S, Miller J. Cosmic radiation shielding tests for UHMWPE fiber/nano-epoxy composites. Composites Science and Technology. 2009;**69**(13):2093-2097

[54] Mani V, Prasad N, Kelkar A. Ultra high molecular weight polyethylene (UHMWPE) fiber epoxy composite hybridized with nanoparticles of gadolinium and boron for radiation shielding. In: Proceedings of SPIE Planetary Defense and Space Environment Applications; 22 September 2016; USA. California: SPIE; 2016. pp. 1-10

[55] Condruz MR, Puscasu C, Voicu LR, Vintila IS, Paraschiv A, Mirea DA. Fiber reinforced composite materials for proton radiation shielding. Materiale Plastic. 2018;**55**(1):5-8

[56] Zegaoui A, Derradji M, Medjahed A, Ghouti HA, Cai W, Liu WB, et al. Exploring the hybrid effects of short glass/basalt fibers on the mechanical, thermal and gamma radiation shielding properties of DCBA/BA-a resin composites. Polymer Plastics Technology and Materials. 2020;**59**(3):311-322